W9-BBS-123

Ecology Basics

Ecology Basics

Volume 2

Mammalian social systems—Zoos
Appendices
Indexes

edited by

The Editors of Salem Press

SALEM PRESS, INC.
Pasadena, California   Hackensack, New Jersey

∞ The paper used in these volumes conforms to the American National Standard for Permanence of Paper for Printed Library Materials,
Z39.48-1992 (R1997).

Library of Congress Cataloging-in-Publication Data
Ecology basics / edited by the editors of Salem Press.
    p. cm. — (Magill's choice)
    Includes bibliographical references.
    ISBN 1-58765-174-2 (set : alk. paper) — ISBN 1-58765-175-0 (v. 1 : alk.
paper) — ISBN 1-58765-176-9 (v. 2 : alk. paper)
    1. Ecology—Encyclopedias. I. Salem Press. II. Series.

QH540.4.E39 2003
577'.03—dc21

2003011370

Second Printing

PRINTED IN THE UNITED STATES OF AMERICA

Contents

Complete List of
Contents

Volume 1

Volume 2

Complete List of Contents

Ecology Basics

MAMMALIAN SOCIAL SYSTEMS

Type of ecology: Behavioral ecology

Social organization in mammals ranges from solitary species, which come together only to breed, to large and intricately organized societies. Understanding the social systems of mammals is essential for effective conservation of species.

All levels of social organization occur in mammals. There are solitary species, such as the mountain lion (*Felis concolor*), in which the male and female adults come together only to mate and the female remains with her young only until they are capable of living independently. At the other numerical extreme are some of the hoofed mammals, which form herds of thousands of individuals. The most socially specialized mammal is probably Africa's naked mole rat (*Heterocephalus glaber*), which has a eusocial colony structure similar to that of ants, bees, and termites. No current theory accounts for the diversity of mammalian social systems, but two broad generalizations are consistently employed to explain mammalian species' social organization. These are the environmental context in which the species exists and the mammalian mode of reproduction.

Reproductive Determinants

More than any other group of animals, mammals are required to form groups for at least part of their lives. Although in all sexually reproducing animals the sexes must come together to mate, mammals have an additional required association between mother and young: All species of mammal feed their young with milk from the mother's mammary glands. This group, a female and her young, is the basis for the development of mammalian social groups. In some species, the social group includes several females and their young and may involve one or more males as well.

Environmental Determinants

Mammalian societies are always organized around one or more females and their offspring. Males may also be part of the group, or they may form separate groups. The size and structure of the group are determined by the ecological setting in which it evolves. The particular ecological factors that seem to be of greatest importance in this determination are food supply, the distribution of the food, and predation (including the hiding places and escape routes available in the habitat).

Large groups occur when food is scattered in a patchy distribution.

These groups are largest when the patches contain abundant food. Many organisms are more likely to find the scattered patches than is a single individual. As long as the patches have enough food for all members of the group, it is to each member's advantage to search with the group. On the other hand, if food is evenly dispersed in small units throughout the environment, the advantage of a group search is lost. Each individual will be better off searching for itself, and some strategy involving a very small social group or even solitary existence would be advantageous.

A somewhat similar argument follows for predators. If large prey are taken, a group of predators should be able to subdue the prey and protect its remains from scavengers more efficiently. If small prey are taken, solitary predators have the advantage, since the prey is easily dispatched and the predator will have it to itself. Many other factors are involved in determining the final form of a species' social organization, but the family unit and environmental context are fundamental in determination of all mammalian social structures.

Primate Social Organization

The primates are the most social group of mammals. Monkeys demonstrate the importance of food supply and its distribution in determining social structure. The olive baboon (*Papio anubis*) occupies savannas, where

Lions are among the few feline species that live in groups, called prides. Each pride consists of only one or a few males and several or many females. The females share the responsibility for raising the cubs, teaching them hunting behaviors and other survival skills. These four littermates are nearly independent and, if male, will likely leave the pride. (PhotoDisc)

it exists in large groups of several adult males, several adult females, and their young. Finding fifty or more animals in a group is not uncommon. Individual males do not guard or try to control specific females except when the females are sexually receptive. The group's food supply is in scattered patches, but each patch contains an abundance of food. The advantage of having many individuals searching for the scattered food is obvious: If any member finds a food-rich patch, there is plenty for all.

Predation probably also plays a role in the olive baboon's social organization. The savannas they roam have many predators and few refuges for escape. A large group is one defense against predators if hiding or climbing out of reach is not practical. Having many observers increases the chance of early detection, giving the prey time to elude the predator. A large group can also mount a more effective defense against a predator. Large groups of baboons use both of these tactics.

The hamadryas baboon (*Papio hamadryas*), on the other hand, lives in deserts in which the food supply is not only scattered but also often found in small patches. The hamadryas baboon's social structure contrasts with that of the olive baboon, perhaps because the small patches do not supply enough food to support large groups. A single adult male, one or a few adult females, and their young make up the basic group of fewer than twenty individuals. Several of these family groups travel together under certain conditions, forming a band of up to sixty animals. Within the band, however, the family groups remain intact. The male of each group herds his females, punishing them if they do not follow him. The bands are probably formed in defense against predators. They break up into family units if predators are absent. At night, hamadryas baboons sleep on cliffs, where they are less accessible to predators. Because suitable cliffs are limited, many family groups gather at these sites. Hundreds of animals may be in the sleeping troop, probably affording further protection against predators.

Though there are exceptions, forest primates consistently live in smaller groups. In many species, fewer than twenty individuals make up the social group at all times. These consist of one or a few mature males, one or a few mature females, and their offspring. The groups are more evenly distributed throughout their habitat than are groups of savanna or desert primates. In forests, the food supply is more abundant and more evenly distributed. Escape from predators is also more readily accomplished—by climbing trees or hiding in the dense cover. Under these conditions, the advantages of large groups are minimal and their disadvantages become apparent. For example, in small groups the competition for mates and food is less.

Ungulate Social Organization

The ungulates have all levels of social organization. African antelope demonstrate social organizations that, in some ways, parallel those of the primates. Forest antelope such as the dik-dik (*Madoqua*) and duiker (*Cephalophus*) are solitary or form small family groups, and they are evenly spaced through their environment. Many hold permanent territories containing the needs of the individual or group. They escape predators by hiding and are browsers, feeding on the leaves and twigs of trees.

Many grassland and savanna antelope, such as wildebeest (*Connochaetes*), on the other hand, occur in large herds. They outrun or present a group defense to predators and are grazers, eating the abundant grasses of their habitat. In many cases, they are also migratory, following the rains about the grasslands to find sufficient food. The social unit is a group of related females and their young. Males leave the group of females and young as they mature. They join a bachelor herd until fully mature, at which time they become solitary, and some establish territories. The large migratory herds are composed of many female-young groups, bachelor herds, and mature males. The social units are maintained in the herd. Though it may seem strange to speak of solitary males in a herd of thousands, that is their social condition. The male territories are permanent in areas that have a reliable food supply year-round, but they cannot be in regions in which the species is migratory. Under these conditions, the males set up temporary breeding territories wherever the herd is located during the breeding season.

There are parallels with primate social patterns. Large groups are formed in grasslands, and these roam widely in search of suitable food. The groups are effective as protection against predators in habitats with few hiding places. Smaller groups are found in forests, where food is more evenly dispersed and places to hide from predators are more readily found.

Rodent Social Organization

Rodents also have all kinds of social organization. The best known, and one of the most complex, is the social system of the black-tailed prairie dog (*Cynomys ludovicianus*). The coterie is the family unit in this case, and it consists of an adult male, several adult females, and their young. Members maintain a group territory defended against members of other coteries. Coterie members maintain and share a burrow system. Elaborate greeting rituals have developed to allow the prairie dogs of a coterie to recognize one another. Hundreds of these coteries occur together in a town. The members of these towns keep the vegetation clipped—as a result, predators can be seen from a distance. Prairie dogs warn one another with a "bark" when they observe a predator, and the burrow system affords a refuge from most predators.

The only vertebrate known to be eusocial is the naked mole rat. It occurs in hot, dry regions of Africa. The colony has a single reproductive female, a group of workers, and a group of males whose only function is to breed with the reproductive female. The workers cooperate in an energetically efficient burrowing chain when enlarging the burrow system. In this way, they are able to extend the burrow system quickly during the brief wet season. Digging is very difficult at other times of the year. The entire social system is thought to be an adaptation to a harsh environment and a sparse food supply.

Carnivore Social Organization

Most carnivores are not particularly social, but some do have elaborate social organization. Many of these are based on the efficiency of group hunting in the pursuit of large prey or on the ability of a group to defend a large food supply from scavengers. The gray wolf (*Canis lupus*) and African hunting dog (*Lycaon pictus*) are examples. In both cases, the social group, or pack, consists of a male and female pair and their offspring of several years. Though there are exceptions, solitary carnivores and carnivores that form temporary family units during the breeding season, such as the red fox (*Vulpes vulpes*), hunt prey smaller than themselves. The coyote (*Canis latrans*) can switch social systems to use the food available most efficiently. It forms packs similar to those of the gray wolf when its main prey is large or when it can scavenge large animals and is solitary when the primary available prey is small.

These examples and many others show that the social groups of mammals are based on the family group. The particular social organization employed by a species is determined by the ecological situation in which it occurs. The specific aspects of the environment that seem to be most important include food abundance, food distribution, food type, and protection from predators.

Carl W. Hoagstrom

See also: Altruism; Communication; Defense mechanisms; Displays; Ethology; Habituation and sensitization; Herbivores; Hierarchies; Insect societies; Isolating mechanisms; Migration; Mimicry; Omnivores; Pheromones; Poisonous animals; Predation; Reproductive strategies; Territoriality and aggression.

Sources for Further Study

Dunbar, Robin I. M. *Primate Social Systems*. Ithaca, N.Y.: Cornell University Press, 1988.

Eisenberg, John F., and Devra G. Kleiman, eds. *Advances in the Study of Mammalian Behavior*. Special Publication 7. Shippensburg, Pa.: American Society of Mammalogists, 1983.

Gittleman, John L., ed. *Carnivore Behavior, Ecology, and Evolution*. Ithaca, N.Y.: Cornell University Press, 1989.

Immelmann, Klaus, ed. *Grzimek's Encyclopedia of Ethology*. New York: Van Nostrand Reinhold, 1977.

Macdonald, David W. *European Mammals: Evolution and Behavior*. London: HarperCollins, 1995.

Nowak, Ronald M., and John L. Paradiso. *Walker's Mammals of the World*. 6th ed. 2 vols. Baltimore: Johns Hopkins University Press, 1999.

Rosenblatt, Jay S., and Charles T. Snowdon, eds. *Parental Care: Evolution, Mechanisms, and Adaptive Significance*. Advances in the Study of Behavior 25. San Diego, Calif.: Academic Press, 1996.

Slater, P. J. B. *An Introduction to Ethology*. Reprint. London: Cambridge University Press, 1990.

Vaughan, Terry A. *Mammalogy*. 4th ed. Philadelphia: Saunders College Publishing, 2000.

Wrangham, Richard W., W. C. McGrew, Frans B. M. De Waal, and Paul G. Heltne, eds. *Chimpanzee Cultures*. Cambridge, Mass.: Harvard University Press, 1994.

MARINE BIOMES

Types of ecology: Aquatic and marine ecology; Biomes; Ecosystem
  ecology

The world's oceans contain the largest and most varied array of life-forms on earth.
The marine environment is divided into coastal, open water, deep-sea, and bottom
zones and the lives of animals living in each of these regions are dictated by the
physical conditions present in these zones.

Approximately 71 percent of earth's surface is covered by salt water,
and the marine environments contained therein constitute the largest
and most diverse array of life on the planet. Life originated in the oceans,
and the salt water that comprises the largest constituent of the tissues of all
living organisms is a vestigial reminder of the aquatic origins of life.

Marine Zones

The marine environment can be divided broadly into different zones, each
of which supports numerous habitats. The coastal area between the high
and low tide boundaries is known as the intertidal zone; beyond this is the
neritic zone, relatively shallow water that extends over the continental
shelves. The much deeper water that extends past the boundaries of the
continental shelves is known as the oceanic zone. Open water of any depth
away from the coastline is also known as the pelagic zone. The benthic
zone is composed of the sediments occurring at the sea floor. Areas in
which freshwater rivers empty into the saltwater oceans produce a continually mixed brackish water region known as an estuary. Estuarine zones
often also include extensive wetland areas such as mudflats or salt
marshes.

   Zones in the marine environment are distributed vertically as well as
horizontally. Life in the ocean, as on land, is ultimately supported by sunlight in most cases, used by photosynthetic plants as an energy source.
Sunlight can only penetrate water to a limited depth, generally between
one hundred and two hundred meters; this region is known as the photic
or epipelagic zone. Below two hundred meters, there may be sufficient
sunlight penetrating to permit vision, but not enough to support photosynthesis; this transitional region may extend to depths of one thousand
meters and is known as the disphotic or mesopelagic zone. Below this
depth, in the aphotic zone, sunlight cannot penetrate and the environment
is perpetually dark, with the exception of small amounts of light produced

by photoluminescent invertebrate and vertebrate animals. This aphotic zone is typically divided into the bathypelagic zone, between seven hundred and one thousand meters as the upper range and two thousand to four thousand meters as the lower range, where the water temperature is between 4 and 10 degrees Celsius. Beneath the bathypelagic zone, overlying the great plains of the ocean basins, is the abyssalpelagic zone, with a lower boundary of approximately six thousand meters. Finally, the deepest waters of the oceanic trenches, which extend to depths of ten thousand meters, constitute the hadalpelagic zone. In each of these zones, the nature and variety of marine life present is dictated by the physical characteristics of the zone. However, these zones are not absolute, but rather merge gradually into each other, and organisms may move back and forth between zones.

Plankton

Marine life can be divided broadly into three major categories. Those small organisms that are either free-floating or weakly swimming and which thus drift with oceanic currents are referred to as plankton. Plankton can be further divided into phytoplankton, which are plantlike and capable of photosynthesis; zooplankton, which are animal-like; and bacterioplankton, which are bacteria and bluegreen algae suspended in the water column. Larger organisms that can swim more powerfully and which can thus move independently of water movements are known collectively as the nekton. Finally, organisms that are restricted to living on or in the sediments of the seafloor bottom are referred to as the benthos.

The phytoplankton, which are necessarily restricted to the photic zone, are by far the largest contributors to photosynthesis in the oceans. The phytoplankton are therefore responsible for trapping most of the solar energy obtained by the ocean (the primary productivity), which can then be transferred to other organisms when the phytoplankton are themselves ingested. The phytoplankton are composed of numerous different types of photosynthetic organisms, including diatoms, which are each encased in a unique "pillbox" shell of transparent silica, and dinoflagellates. The very rapid growth of some species of dinoflagellates in some areas results in massive concentrations or blooms that are sometimes referred to as red tides. Chemicals that are produced by red tide dinoflagellates often prove toxic to other marine organisms and can result in massive die-offs of marine life. Smaller photosynthetic plankton forms comprise the nanoplankton and also play an important role in the photosynthetic harnessing of energy in the oceans.

The zooplankton are an extremely diverse group of small animal organisms. Unlike the phytoplankton, which can make their own complex organic compounds via photosynthesis, the phytoplankton must ingest or absorb organic compounds produced by other organisms. This is accomplished by either preying upon other planktonic organisms or by feeding on the decaying remains of dead organisms. A number of zooplankton species also exist as parasites during some portion of their life cycles, living in or upon the bodies of nekton species. The largest group of zooplankton are members of the subphylum Crustacea, especially the copepods. These organisms typically possess a jointed exoskeleton, or shell, made of chitin, large antennae, and a number of jointed appendages. Space precludes a definitive listing of all of the zooplanktonic organisms, however virtually all of the other groups of aquatic invertebrates are represented in the bewildering variety of the zooplankton, either in larval or adult forms. Even fish, normally a part of the nekton, contribute to the zooplankton, both as eggs and as larval forms.

The bacterioplankton are found in all of the world's oceans. Some of these, the blue-green algae (cyanobacteria), play an important role in the photosynthetic productivity of the ocean. Bacterioplankton are usually found in greatest concentrations in surface waters, often in association with organic fragments known as particulate organic carbon, or marine snow. Bacterioplankton play an important role in renewing nutrients in the photic zones of the ocean; such renewal is important in maintaining the photosynthetic activity of the phytoplankton, upon which the rest of marine life is in turn dependent.

One of the principal problems facing plankton is maintaining their position in the water column. Since these organisms are slightly denser than the surrounding seawater, they tend to sink. Clearly this is a disadvantage, particularly since plankton typically have very limited mobility. This is especially true for the photosynthetic phytoplankton, which must remain within the photic zone in order to carry on photosynthesis. A number of strategies have evolved among planktonic species to oppose this tendency to sink. Long, spindly extensions of the body provide resistance to the flow of water. Inclusions of oils or fats (which are less dense than water) within the body provide positive buoyancy by decreasing the overall density of the plankton. Finally, some species, such as the Portuguese man-o'-war, generate balloonlike gas bladders, which provide enough buoyancy to keep them at the very surface of the epipelagic zone.

Nekton

The nekton comprise those larger animals that have developed locomotion to a sufficient degree that they can move independently of the ocean's water movements. Whereas the plankton are principally invertebrates, most of the nekton are vertebrates. The majority of the nekton are fish, although reptile, bird, and mammalian species are also constituent parts. The oceanic nekton are those species which are found in the epipelagic zone of the open ocean. These include a wide variety of sharks, rays, bony fish, seabirds, marine mammals, and a few species of reptiles. Some members of the oceanic nekton, such as blue sharks, oceanic whitetip sharks, tuna, flying fish, and swordfish, spend their entire lives in the pelagic environment; these are said to be holoepipelagic. Others, the meroepipelagic nekton—such as herring, dolphins, salmon, and sturgeon—spend only a portion of their lives in the epipelagic zone, returning to coastal or freshwater areas to mate.

Seabirds are a special case: Although they spend much of their time flying over the epipelagic zone and nest on land, they feed in the epipelagic zone. Some species may dive as deep as one hundred meters in search of prey. Some members of the nekton enter the epipelagic only at certain times in their life cycles. Eels of the family *Anguillidae* spend most of their lives in fresh water but return to the epipelagic zone to spawn. Additionally, at night many species of deep-water fish migrate up into the epipelagic to feed before returning to deeper waters during the daylight hours.

The pelagic environment, unlike the terrestrial one, is profoundly three-dimensional. Nektonic animals can move both horizontally and vertically within the water column. Furthermore, since most of the pelagic environment is essentially bottomless, since there is no apparent or visible ground or substrate, the environment is essentially uniform and featureless. These features play an important role in the evolution of the behavior of nektonic animals. Fish suspended in an essentially transparent and featureless medium have no shelter in which to hide from predators, nor are there any apparent landmarks to serve as directional cues for animals moving horizontally from place to place. Life in the open ocean has therefore favored adaptations for great mobility and speed with which to move across large distances and escape from predators, as well as camouflage and cryptic coloration designed to deceive potential predators or prey.

As is the case for plankton, most nektonic animals are denser than the surrounding seawater, and maintaining position in the water column is of the first importance. Most fish possess a swim bladder, a gas-filled membranous sac within their body that opposes the tendency to sink and provides the fish with neutral buoyancy. Sharks and rays lack a swim bladder, but accumulate large concentrations of fats and oils in their liver, which also help counter the tendency to sink. Large, fast-swimming species of

Seabirds and fish are examples of nekton, generally vertebrates that have developed locomotion to a sufficient degree that they can move independently of the ocean's water movements or occupy other portions of the photic, or epipelagic, zone of the ocean ecosystem. (PhotoDisc)

shark, tuna, and many billfish also rely on the generation of hydrodynamic lift to maintain vertical position in the water column. The tail and body of these fish generate forward thrust, moving the animal through the water, and the fins, notably the pectoral fins, generate lift from the water flowing over them in a manner similar to that of an airplane's wing. Thus these animals fly through the water, but are in turn required to move continuously in order to generate lift.

All members of the nekton are carnivores, feeding on other nektonic species or upon plankton, particularly the larger zooplankton. In general, the size of the prey consumed by nekton is directly related to the size of the predator, with larger species consuming larger prey. However, the organisms that feed upon plankton, the planktivores, include a wide variety of fish species such as herring, salmon, and the whale shark, the largest extant fish species. They also include the largest marine animals of all, the baleen whales. The case of large animals feeding upon very small plankton directly addresses the need of all animals to meet their energy requirements. For all animals, the amount of energy obtained from food consumed must necessarily exceed the energy expended in acquiring the prey. Very large animals, such as whales and whale sharks, require a great deal of energy to move their bodies through the aquatic environment, but because of their great size they are necessarily less agile than smaller forms. The amount of energy required to chase and catch these smaller animals would generally exceed the energy derived from ingesting them. Plankton, however, are relatively easy to obtain due to their very limited mobility. However, because of their small size, vast quantities of plankton must be ingested in order to meet the metabolic requirements of large marine animals. Some very large species that are not planktivores solve the energy problem by evolving behaviors for acquiring specialized diets that yield higher energy. White sharks, for example, feed on fish when young, but as they age and increase in size, marine mammals, notably seals and sea lions (pinnipeds), become a major part of their diet. Marine mammals all possess blubber, an energy-rich substance that yields much more energy than fish. Similarly, sperm whales, the largest hunting carnivores on the planet, have a diet that consists in large part of giant squid, which are hunted in the ocean depths largely using the whale's acoustic echolocation sense. Orcas (killer whales) effectively use pack hunting techniques to hunt larger whales and other marine mammals.

The deeper regions of the ocean are dominated by different types of nekton. However, we know even less about their ecology due to their relative inaccessibility. The disphotic or mesopelagic zone contains many animal species that migrate vertically into surface waters at night to feed upon the plankton there. Many of these organisms possess large, well-developed eyes and also possess light organs containing symbiotic luminescent bacteria. The majority of the fish species in this group are colored black and the invertebrates are largely red (red light penetrates water less effectively than do longer wavelengths, and these animals appear dark-colored at depth). Beneath this zone, in the bathypelagic and abyssalpelagic zones, there are many fewer organisms and much less diversity than in the shallower levels. Animals in this region are typically colorless and possess small eyes and luminescent organs. Because organisms in these deep regions are few and far between, many species have become specialized in order to maximize their advantages. Thus, deep-sea fish are characterized by large teeth and remarkably hinged jaws that allow them to consume prey much larger than might be expected from their size. Similarly, since encounters with potential mates are presumably scarce, a number of unique reproductive strategies have evolved. In the anglerfish (*Ceratius*), all of the large individuals are female and the comparatively tiny males are parasitic, permanently attaching themselves to the female. Much, however still remains to be learned of the ecology of these deep-sea organisms.

Benthos

The benthos of the world's oceans consists of animals that live on the solid substrate of the water column, the ocean floor. Scientists typically divide benthic organisms into two categories, the epifauna, which live on the surface of the bottom at the sediment-water interface and the infauna, those organisms living within the sediments. In shallow water benthic communities, members of virtually every major animal group are represented. Ecologists generally differentiate between soft bottom benthic communities (sand, silt, and mud, which comprise the majority of the benthic zone) and rocky bottom communities, which are less common proportionately. Soft bottom communities have an extensive diversity of burrowing infauna, such as polychaete worms, and mollusks, such as clams. Rocky bottom communities possess a larger proportion of epifauna, such as crustaceans and echinoderms (starfish, sea urchins, and brittle stars), living on the surface of what is essentially a two-dimensional environment. Vertical faces of the hard bottom environment, such as canyon walls or coral reefs, are often home to a wide variety of animals occupying various crannies and caves. In some parts of the world, kelp plants that are anchored to the substrate and which extend to the water surface dominate the rocky bottom substrate. In these kelp forests, large kelp plants (actually a species of brown algae) form a forestlike canopy that plays host to a wide and complex array of animals extending throughout the water column. On the deep ocean floor, the benthos is composed of representatives of virtually major animal group: crustaceans such as amphipods, segmented polychaete worms, sea cucumbers, and brittle stars. Less common are starfish, sea lilies, anemones, and sea fans. The fish of the deep benthos include rat tails and a number of eel species.

Estuaries, where freshwater rivers empty into marine environments, are typified by large, cyclic changes in temperature and salinity. Although estuaries have played an important role in human history as the sites of major ports, the variety and number of estuarine species tend to show less diversity of animal species due to the difficulty in adapting to the large swings in environmental conditions.

Animal life in the sea, like that on land, shows an astonishing variety of forms and behaviors, the result of natural selection. The inaccessibility and hostility of much of the world's oceans to human exploration and observation leaves much yet to be learned about the biology of marine life. Much remains to be achieved in order to obtain a useful body of knowledge concerning life in the sea.

John G. New

Marine biomes

See also: Acid deposition; Eutrophication; Evolution: definition and theories; Habitats and biomes; Invasive plants; Lakes and limnology; Ocean pollution and oil spills; Reefs; Wetlands.

Sources for Further Study

Niesen, T. M. *The Marine Biology Coloring Book.* 2d ed. New York: HarperResource, 2000.
Nybakken, J. W. *Marine Biology: An Ecological Approach.* 5th ed. San Francisco: Benjamin/Cummings, 2001.
Robison, B. H., and J. Connor. *The Deep Sea.* Monterey Bay, Calif.: Monterey Bay Aquarium Press, 1999.
Safina, C. *Song for the Blue Ocean: Encounters Along the World's Coasts and Beneath the Seas.* New York: Henry Holt, 1998.

MEDITERRANEAN SCRUB

Types of ecology: Biomes; Ecosystem ecology

Mediterranean scrub vegetation is dominated by fire-adapted shrubs. The biome fringes the Mediterranean Sea, for which it is named, but is also found along western coasts of continents in areas with warm, dry summers and moist, cool winters.

Regions with mediterranean vegetation are coastal regions between 30 and 45 degrees north latitude or between 30 and 45 degrees south latitude. The air circulating around high-pressure zones over adjacent oceans guides storms away from the coast in the warm season but changes position in concert with the tilt of the earth on its axis and brings storms onto the coast in the cool season. As a result, the warm season is dry, and the cool season is moist. Fire is an important component of mediterranean environments, especially after the warm, dry summer.

North America's representative of mediterranean scrub is the chaparral of the Pacific Coast of Southern California and northern Baja California, Mexico. In chaparral and some other mediterranean regions, winds blowing from continental high-pressure regions toward the coast help push storm tracks offshore during the warm season. In California these winds are called Santa Ana winds and are best known for driving chaparral fires. Lightning started such fires before human settlement, but they are often started by careless people today. With the cooler temperatures of autumn and winter the continental pressure wanes and the Santa Ana winds decrease. At the same time, the oceanic high-pressure region shifts, and winter storms track onto the coast, bringing the cool season rains.

Character and Components

Mediterranean scrub is found in small, scattered areas around the world. The plant species that occur in this biome on one continent are unrelated to those that occur in the same biome on other continents. As a result, mediterranean scrub presents a classical example of convergent evolution, the environmentally driven development of similar characteristics in unrelated species. Under the influence of mediterranean climate, entire communities of unrelated species become similar to one another. Many mediterranean areas also contain a large number of endemic plant species, species that grow nowhere else.

Mediterranean scrub is dominated by shrubs well adapted to fire. Some species have specialized underground structures that are undamaged by the fire and send up new growth shortly after the fire passes. Other species have specialized, long-lived seeds that require intense heat to stimulate germination. Still other species combine the two strategies. In communities that burn regularly, such species have a great advantage over their competitors.

Mediterranean shrubs are not just adapted to recover after a fire; they are actually adapted to carry the fire once it is started. These species synthesize and store highly flammable chemicals in their leaves and stems. The flammable vegetation ensures that most fires will burn large areas.

The most widespread shrub in North American chaparral is chamise (*Adenostoma fasciculatum*), which sprouts from underground structures and produces large numbers of seedlings after a fire. Various species of manzanita (*Arctostaphylos*) and wild lilac (*Ceanothus*) are also widespread throughout chaparral. Some species in each genus both sprout and produce large numbers of seedlings after fires. Other species in each genus depend entirely on heat-stimulated seeds to reestablish their presence in a burned area.

Mediterranean vegetation also occurs on western coasts in southern Australia, where it is called mallee; the Cape region of South Africa (fynbos); the central coast of Chile (matorral); and around the Mediterranean Sea (maquis). In all these areas, the vegetation has the same adaptive characteristics and appearance, but the species are not related to those of other areas. Although there are differences among the regions besides the species that occur in each, the similar physical and vegetational characteristics lend a continuity that is widely recognized as the mediterranean scrub biome.

Human Impact

As people moved into Mediterranean scrub regions, two major and related concerns surfaced. First, the fires, which are such an important part of scrub ecology, were destructive and dangerous, leading to fire suppression. Second, fire suppression may actually increase fire damage and may threaten the mediterranean scrub biome's very existence when combined with other human activities. A comparison of the fire history in the chaparral of California and that of Baja California lends credibility to the idea that fire suppression increases fire damage. Fire suppression has long been practiced in Southern California. In contrast, much less fire suppression has gone on in Baja. Fewer, larger, and more destructive fires burn in Southern California chaparral than in Baja chaparral. The simplest explanation is that fire suppression allows fuel to build up, so that when a fire starts it is essentially unstoppable, as often occurs in California chaparral.

With less fire suppression and less fuel accumulation, Baja fires burn more frequently but are smaller and less destructive. The small fires remove the fuel periodically, thus decreasing the danger of large, destructive fires.

There are other differences between California and Baja chaparral that may account for the differences in the fire regimes, but the foregoing hypothesis is interesting from the perspective of human impact on chaparral as well as that of fire's impact on humans. Population growth and its attendant activities threaten the very existence of the chaparral. Humans destroy chaparral to build home sites, suppress fires, and plant grass in burned areas to stabilize the soil and to mitigate future fires. The grasses compete with chaparral plants and retard chaparral recovery. The impact of these and other activities on the native chaparral ecosystem is not well understood but is almost certainly negative. Other mediterranean scrub areas suffer similar fates. Although mediterranean scrub is still well represented in comparison to some biomes, its response to human impact should be carefully studied and monitored, both to protect human investment in mediterranean ecosystems and to preserve the intriguing mediterranean scrub and its many unique plant species.

Carl W. Hoagstrom

See also: Biomes: determinants; Biomes: types; Chaparral; Forest fires; Forest management; Forests.

Sources for Further Study

Barbour, Michael G., and William Dwight Billings, eds. *North American Terrestrial Vegetation*. 2d ed. New York: Cambridge University Press, 2000.

Dallman, Peter R. *Plant Life in the World's Mediterranean Climates*. Berkeley: University of California Press, 1998.

Vankat, John L. *The Natural Vegetation of North America: An Introduction*. Melbourne, Fla.: Krieger, 1992.

METABOLITES

Type of ecology: Chemical ecology

Metabolites are compounds synthesized by plants for both essential functions, such as growth and development (primary metabolites), and specific functions, such as pollinator attraction of defense against herbivory (secondary metabolites).

Metabolites are organic compounds synthesized by organisms using enzyme-mediated chemical reactions called metabolic pathways. Primary metabolites have functions that are essential to growth and development and are therefore present in all plants. In contrast, secondary metabolites are variously distributed in the plant kingdom, and their functions are specific to the plants in which they are found. Secondary metabolites are often colored, fragrant, or flavorful compounds, and they typically mediate the interaction of plants with other organisms. Such interactions include those of plant-pollinator, plant-pathogen, and plant-herbivore.

Primary Metabolites

Primary metabolites comprise many different types of organic compounds, including, but not limited to, carbohydrates, lipids, proteins, and nucleic acids. They are found universally in the plant kingdom because they are the components or products of fundamental metabolic pathways or cycles such as glycolysis, the Krebs cycle, and the Calvin cycle. Because of the importance of these and other primary pathways in enabling a plant to synthesize, assimilate, and degrade organic compounds, primary metabolites are essential.

Examples of primary metabolites include energy-rich fuel molecules, such as sucrose and starch, structural components such as cellulose, informational molecules such as DNA (deoxyribonucleic acid) and RNA (ribonucleic acid), and pigments, such as chlorophyll. In addition to having fundamental roles in plant growth and development, some primary metabolites are precursors (starting materials) for the synthesis of secondary metabolites.

Secondary Metabolites

Secondary metabolites largely fall into three classes of compounds: alkaloids, terpenoids, and phenolics. However, these classes of compounds also include primary metabolites, so whether a compound is a primary or secondary metabolite is a distinction based not only on its chemical structure but also on its function and distribution within the plant kingdom.

Many thousands of secondary metabolites have been isolated from plants, and many of them have powerful physiological effects in humans and are used as medicines. It is only since the late twentieth century that secondary metabolites have been clearly recognized as having important functions in plants. Research has focused on the role of secondary metabolites in plant defense. This is discussed below with reference to alkaloids, though it is relevant to many types of secondary metabolites.

Alkaloids

Alkaloids are a large group of nitrogen-containing compounds, examples of which are known to occur in approximately 20 percent of all flowering plants. Closely related plant species often contain alkaloids of related chemical structure. The primary metabolites from which they are derived include amino acids such as tryptophan, tyrosine, and lysine. Alkaloid biosynthetic pathways can be long, and many alkaloids have correspondingly complex chemical structures. Alkaloids accumulate in plant organs such as leaves or fruits and are ingested by animals that consume those plant parts. Many alkaloids are extremely toxic, especially to mammals, and act as potent nerve poisons, enzyme inhibitors, or membrane transport inhibitors. In addition to being toxic, many alkaloids are also bitter or otherwise bad-tasting. Therefore, the presence of alkaloids and other toxic secondary metabolites can serve as a deterrent to animals to avoid eating such plants.

Sometimes domesticated animals that have not previously been exposed to alkaloid-containing plants do not have acquired avoidance mechanisms, and they become poisoned. For example, groundsel contains the alkaloid senecionine, which has resulted in many recorded cases of livestock fatalities due to liver failure. More frequently, over time, natural selection has resulted in animals developing biochemical mechanisms or behavioral traits that lead to avoidance of alkaloid-containing plants.

In other, more unusual cases, animals may evolve a mechanism for sequestering (storing) or breaking down a potentially toxic compound, thus "disarming" the plant. For instance, caterpillars of the cinnabar moth can devour groundsel plants and sequester senecionine without suffering any ill effects. Moreover, the caterpillars thereby acquire their own weapon against predators: the plant-derived alkaloid stored within their bodies. Over time, plants acquire new capabilities to synthesize additional defense compounds to combat animals that have developed "resistance" to the original chemicals. This type of an "arms race" is a form of coevolution and may help to account for the incredible abundance of secondary metabolites in flowering plants.

Medicinal Alkaloids

Many potentially toxic plant-derived alkaloids have medicinal properties, as long as they are administered in carefully regulated doses. Alkaloids with important medicinal uses include morphine and codeine from the opium poppy and cocaine from the coca plant. These alkaloids act on the nervous system and are used as painkillers. Atropine, from the deadly nightshade plant, also acts on the nervous system and is used in anesthesia and ophthalmology. Vincristine and vinblastine from the periwinkle plant are inhibitors of cell division and are used to treat cancers of the blood and lymphatic systems. Quinine from the bark of the cinchona tree is toxic to the Plasmodium parasite, which causes malaria, and has long been used in tropical and subtropical regions of the world. Other alkaloids are used as stimulants, including caffeine, present in coffee, tea, and cola plants (and the drinks derived from these plants); and nicotine, which is present in tobacco. Nicotine preparations are, paradoxically, also used as an aid in smoking cessation. Nicotine is also a very potent insecticide. For many years ground-up tobacco leaves were used for insect control, but this practice was superseded by the use of special formulations of nicotine. More recently the use of nicotine as an insecticide has been discouraged because of its toxicity to humans.

Terpenoids

Terpenoids are derived from acetyl coenzyme A or from intermediates in glycolysis. They are classified by the number of five-carbon isoprenoid units they contain. Monoterpenes (containing two C_5-units) are exemplified by the aromatic oils (such as menthol) contained in the leaves of members of the mint family. In addition to giving these plants their characteristic taste and fragrance, these aromatic oils have insect-repellent qualities. The pyrethroids, which are monoterpene esters from the flowers of chrysanthemum and related species, are used commercially as insecticides. They fatally affect the nervous systems of insects while being biodegradable and nontoxic to mammals, including humans.

Diterpenes are formed from four C_5-units. Paclitaxel (commonly known by the name Taxol), a diterpene found in bark of the Pacific yew tree, is a potent inhibitor of cell division in animals. At the end of the twentieth century, paclitaxel was developed as a powerful new chemotherapeutic treatment for people with solid tumors, such as ovarian cancer patients.

Triterpenoids (formed from six C_5-units) comprise the plant steroids, some of which act as plant hormones. These also can protect plants from insect attack, though their mode of action is quite different from that of the pyrethroids. For example, the phytoecdysones are a group of plant sterols that resemble insect molting hormones. When ingested in excess, phytoecdysones can disrupt the normal molting cycle with often lethal consequences to the insect.

Tetraterpenoids (eight C_5-units) include important pigments such as beta-carotene, which is a precursor of vitamin A, and lycopene, which gives tomatoes their red color. Rather than functioning in plant defense, the colored pigments that accumulate in ripening fruits can serve as attractants to animals, which actually aid the plant in seed dispersal.

The polyterpenes are polymers that may contain several thousand isoprenoid units. Rubber, a polyterpene in the latex of rubber trees that probably aids in wound healing in the plant, is also very important for the manufacture of tires and other products.

Phenolic Compounds

Phenolic compounds are defined by the presence of one or more aromatic rings bearing a hydroxyl functional group. Many are synthesized from the amino acid phenylalanine. Simple phenolic compounds, such as salicylic acid, can be important in defense against fungal pathogens. Salicylic acid concentration increases in the leaves of certain plants in response to fungal attack and enables the plant to mount a complex defense response. Interestingly, aspirin, a derivative of salicylic acid, is routinely used in humans to reduce inflammation, pain, and fever. Other phenolic compounds, called isoflavones, are synthesized rapidly in plants of the legume family when they are attacked by bacterial or fungal pathogens, and they have strong antimicrobial activity.

Lignin, a complex phenolic macromolecule, is laid down in plant secondary cell walls and is the main component of wood. It is a very important structural molecule in all woody plants, allowing them to achieve height, girth, and longevity. Lignin is also valuable for plant defense: Plant parts containing cells with lignified walls are much less palatable to insects and other animals than are nonwoody plants and are much less easily digested by fungal enzymes than plant parts that contain only cells with primary cellulose walls.

Other phenolics function as attractants. Anthocyanins and anthocyanidins are phenolic pigments that impart pink and purple colors to flowers and fruits. This pigmentation attracts insects and other animals that move between individual plants and accomplish pollination and fruit dispersal. Often the plant pigment and the pollinator's visual systems are well matched: Plants with red flowers attract birds and mammals because these animals possess the correct photoreceptors to see red pigments.

Valerie M. Sponsel

See also: Allelopathy; Defense mechanisms; Genetically modified foods; Pheromones; Poisonous animals; Poisonous plants.

Sources for Further Study
Levetin, Estelle, and Karen McMahon. *Plants and Society*. Boston: WCB/ McGraw-Hill, 1999.
Moore, Randy, et al. *Botany*. 2d ed. Boston: WCB/McGraw-Hill, 1998.

MIGRATION

Type of ecology: Behavioral ecology

Migration is an important ecological process that results in the redistribution of animal populations from one habitat to another. Adaptive habitat changes are fundamentally important aspects of the life histories of many species.

Migration is a general term employed by ecologists and ethologists to describe the nearly simultaneous movement of many individuals or entire populations of animals to or between different habitats. As defined, migrations do not include local excursions made by individuals or small groups of animals in search of food, to mark territorial boundaries, or to explore surrounding environments.

Nomads are migrants whose populations follow those of their primary food sources. Such animals (the American bison, for example) do not have fixed home ranges and wander in search of suitable forage. Some scientists view nomadic movements as a form of extended foraging behavior rather than as a special case of migration. In either context, the important point is that populations change habitats in response to changing conditions.

In contrast to migrations made by populations and excursions made by individuals, the spreading or movement of animals away from others is known as dispersal. Examples of dispersal include the drift of plankton in currents and the departure of subadult animals from the home range of their parents. In numerous species (sea turtles, rattlesnakes, and salmon, for example), dispersed members of a population may return to the place of origin after a variable interval of time.

Navigation

Among the animals known to navigate are birds (the best-studied group), lobsters, bees, tortoises, bats, marine and terrestrial mammals, fish, brittle starfish, newts, toads, and insects. Some migratory species can orient themselves—that is, they know where they are in time and space. Many birds and mammals, for example, have an inherent sense of the direction, distance, and location of distant habitats. Orientation and travel along unfamiliar routes from one place or habitat to another is called navigation. Navigators use environmental and sensory information to reach distant geographical locations, and many of them do so with a remarkably accurate sense of timing. Homing pigeons are perhaps the best-studied animal navigators. These birds are able not only to discover where they are when released but also to return to their home loft from distant geographical locations.

Much has been learned about how animals successfully navigate over long distances from the pioneering studies of Archie Carr. Carr proposed that green sea turtles successfully find their widely separated nesting and feeding beaches by means of an inherent clock sense, map sense, and compass sense. His investigations and those of many others continue to stimulate great interest in the physiology and ecology of navigating species and in the environmental cues to which they respond. Sensory biologists, biophysicists, and engineers have incorporated knowledge of how animals detect and use environmental information to develop new and more accurate navigational systems for human use.

Animals use a variety of cues to locate their positions and appropriate travel paths. Most species have been found to use more than one type of information (sequentially, alternatively, or simultaneously) to navigate. Included among the orientation guideposts that one or more of these groups may use are the positions of the sun and stars, magnetic fields, ultraviolet light, tidal fluctuations caused by the changing positions of the moon and sun, atmospheric pressure variations, infrasounds (very low frequency sounds), polarized light (on overcast days), environmental odors, shoreline configurations, water currents, and visual landmarks. Celestial cues also require a time sense, or an internal clock, to compensate for movements of the animal relative to changing positions of celestial objects in the sky. In addition to an absolute dependence on environmental cues, young or inexperienced members of some species may learn navigational routes from experienced individuals, such as their parents, or other experienced individuals in the population. Visual mapping remembered from exploratory excursions may also play a role in enhancing the navigational abilities of some birds, fish, mammals, and other animals.

Benefits of Migration

The different categories of animal movements, however, are perhaps not as important as the reasons animals migrate and the important biological consequences of the phenomenon. As a general principle, migrations are adaptive behavioral responses to changes in ecological conditions. Populations benefit in some way by regularly or episodically moving from one habitat to another.

An example of the adaptive value of migratory behavior is illustrated by movement of a population from a habitat where food, water, space, nesting materials, or other resources have become scarce (often a seasonal phenomenon) to an area where resources are more abundant. Relocation to a new habitat (or to the same type of habitat in a different geographical area) may reduce intraspecific or interspecific competition, may reduce death rates, and may increase overall fitness in the population. These benefits may result in an increase in reproduction in the population. Reproductive success, then, is the significant benefit and the only biological criterion used to evaluate population fitness.

Programmed Movements

While many factors are believed to initiate migratory events, most fall into one of two general categories. The first and largest category may be called programmed movements. Such migrations usually occur at predictable intervals and are important characteristics of a species' lifestyle or life cycle. Programmed migrations are not, in general, density-dependent. Movements are not caused by overcrowding or other stresses resulting from an excessive number of individuals in the population.

The lifestyle of a majority of drifting animals whose entire lives are spent in the water column, for example, includes a vertical migration from deep water during the day to surface waters at night. Thus, plankton exhibit a circadian rhythm (activity occurring during twenty-four-hour intervals) in their movements. An abundance of food at or near the surface and escape from deep-water predators are among the possible reasons for these migrations. Daily vertical movements of plankton are probably initiated by changes in light intensity at depth, and the animals follow light levels as they move toward the surface with the sinking sun. It is interesting to note that zooplankters living in polar waters during the winter-long night do not migrate.

Monarch butterflies and many large, vertebrate animals, such as herring, albatross, wildebeests, and temperate-latitude bats, migrate from one foraging area to another, or from breeding to foraging habitats, on a seasonal or annual basis. Annual migrations usually coincide with seasonal variation. Changes in day length, temperature, or the abundance of preferred food items associated with seasonal change may stimulate mass movements directly, or indirectly, through hormonal or other physiological changes that are correlated with seasonal environmental change. The onset of migration in many vertebrates is evidenced by an increase in restlessness that seems, in human terms, to be anticipatory.

In addition to their daily vertical migrations (lifestyle movements), the life cycles of marine zooplankton involve migrations, and it is convenient to use them as examples. As discussed, most adult animal plankters are found at depth during the day and near the surface at night. In contrast, zooplankton eggs and larvae remain in surface waters both day and night.

Bird migrations are among the most noticeable of animal migrations, with huge flocks passing through the skies on their way to a seasonal home that can be thousands of miles distant from their original location. (Corbis)

As the young stages grow, molt, and change their shapes and food sources, they begin to migrate vertically. The extent of vertical migrations gradually increases throughout the developmental period, and as adults, these animals assume the migratory patterns of their parents. Patterns of movement that change during growth and development are examples of ontogenetic, or life-cycle, migrations.

Episodic Movements

The second large category of migratory behavior includes episodic, density-dependent population movement. Such migrations are often associated with, or caused by, adverse environmental changes (effect) that may be caused by overlarge populations (cause). Local resources are adequate to support a limited number of individuals (called the carrying capacity of the environment), but once that number has been exceeded, the population must either move or perish. Unfortunately, migration to escape unfavorable conditions may be unsuccessful, as another suitable habitat may not be encountered. Migrations caused by overpopulation or environmental degradation are common. Pollution and habitat destruction by humankind's activities are increasingly the cause of degraded environments, and in such cases, it is reasonable to conclude that humans have reduced the carrying capacity of many animal habitats. Familiar examples of density-dependent migrations are those of lemmings, locusts, and humans.

Ecological Import

Environmental or physiological factors that initiate migrations may be of interest to sensory biologists and physiological ecologists; knowledge of variation in population distributions is important to biogeographers and wildlife biologists; and migrations in predator-prey relationships, competition, pollution, and life-history strategies are important aspects of classical ecological studies.

In addition to the specific aspect of migration being studied, the particular group of animals under investigation (moths, eels, elephants, snails) requires that different methods be used. Some of the approaches used in migration-related research illustrate how information and answers are obtained by scientists.

Arctic terns migrate from their breeding grounds in the Arctic to the Antarctic pack ice each year. The knowledge that these birds make a twenty-thousand-mile annual round-trip comes from the simplest and most practical method: direct observation of the birds (or their absence) at either end of the trip. Direct observation by ornithologists of the birds in flight can establish what route they take and whether they pause to rest or feed en route. Many birds have also been tracked using radar or by observations of their silhouettes passing in front of the moon at night. Birds are often banded (a loose ring containing coded information is placed on one leg) to determine the frequency of migration and how many round-trips an average individual makes during its lifetime. From this information, estimates of longevity, survivorship rates, and nesting or feeding site preferences can be made.

Factors that initiate migratory behavior in terns and in other birds can often be determined by ecologists able to relate environmental conditions (changes in temperature, day length, and the like) to the timing of migrations. Physiological ecologists study hormonal or other physiological changes that co-occur with environmental changes. Elevated testosterone levels, for example, may signal the onset of migratory behavior.

How Arctic terns orient and navigate along their migratory routes is usually studied by means of laboratory-conducted behavioral experiments. Birds are exposed to various combinations of stimuli (magnetic fields, planetarium-like celestial fields, light levels), and their orientation, activity levels, and physiological states are measured. Experiments involving surgical or chemical manipulation of known sensory systems are sometimes conducted to compare behavioral reactions to experimental stimuli. In such experiments, the birds (or other test animals) are rarely harmed.

Tags of several types are used to study migrations in a wide variety of animals, including birds, bees, starfish, reptiles, mammals, fish, snails, and many others. Tags may be transmitting collars (located by direction-finding radio receivers); plastic or metal devices attached to ears, fins, or flippers; or even numbers, painted on the hard exoskeletons of bees and other insects. Additional types of tagging (or identifying) include radioactive implants and microchips that can be read by computerized digitizers; the use of brands and tattoos; and, of great interest, the use of biological tags. Parasites known to occur in only one population of migrants (nematode parasites of herring, for example) provide an interesting illustration of how the distribution of one species can be used to provide information about another.

The Importance of Migration

The causes, frequency, and extent of animal migration are so diverse that several definitions for the phenomenon have been proposed. None of these has been accepted by all scientists who study animal movements, however, and it is sometimes difficult to interpret what is meant when the term "migration" is used. Most researchers have adopted a broad compromise to include all but trivial population movements that involve some degree of habitat change.

It is important to recognize that few populations of animals are static; even sessile animals (such as oysters and barnacles) undergo developmental habitat changes, which are referred to as ontogenetic migrations. Aside from certain tropical and evergreen forest areas where migrations are relatively uncommon, a significant number of both aquatic and terrestrial species move from one habitat to another at some time during their lives. In the face of environmental change, including natural events such as seasonal variation and changes caused by resource limitations and environmental degradation, animals must either move, perish, or escape by means of drastic population reduction or by becoming inactive until conditions become more favorable (hibernation, arrested development and dormancy, and diapause in insects are examples of behavioral-ecological inactivity). Migration is the most common behavioral reaction to unfavorable environmental change exhibited by animals.

One cannot understand the biology of migrators until their distribution and habitats throughout life are known. The patterns of animal movements are fascinating, and it is useful to summarize some of the major differences between them. First, many species travel repeatedly during their lives between two habitats, on a daily basis (as plankton and chimney swifts do) or on an annual basis (as frogs and elks do). Second, some species migrate from one habitat (usually suitable for young stages) to another (usually the adult habitat) only once during their lives (for example, salmon, eels, damselflies, and most zooplankton, which live on the bottom as adults). Third, some species (many butterflies, for example) are born and mature in one geographical area (England, for example), migrate as adults to a distant geographical area (Spain, for example), and produce offspring that mature in the second area. These migrations take place between generations. In a fourth pattern, one may include the seasonal swarming of social arthropods such as termites, fire ants, and bees. A fifth but ill-defined pattern is discernible, exemplified by locust "plagues," irruptive emigration in lemmings and certain other rodents, and some mass migrations by humankind, as caused by war, famine, fear, politics, or disease. These are episodic and often, if not primarily, caused by severe population stress or catastrophic environmental change.

Sneed B. Collard

See also: Altruism; Communication; Defense mechanisms; Displays; Ethology; Habituation and sensitization; Herbivores; Hierarchies; Insect societies; Isolating mechanisms; Mammalian social systems; Mimicry; Omnivores; Pheromones; Poisonous animals; Predation; Reproductive strategies; Territoriality and aggression.

Sources for Further Study

Able, Kenneth P., ed. *Gatherings of Angels: Migrating Birds and Their Ecology.* Ithaca, N.Y.: Comstock, 1999.

Aidley, David, ed. *Animal Migration.* New York: Cambridge University Press, 1981.

Begon, Michael, John Harper, and Colin Townsend. *Ecology: Individuals, Populations, and Communities.* 3d ed. Sunderland, Mass.: Sinauer Associates, 1996.

Dingle, Hugh. *Migration: The Biology of Life on the Move.* New York: Oxford University Press, 1996.

Eisner, Thomas, and Edward Wilson, eds. *Animal Behavior: Readings from "Scientific American."* San Francisco: W. H. Freeman, 1975.

Newberry, Andrew, ed. *Life in the Sea: Readings from "Scientific American."* San Francisco: W. H. Freeman, 1982.

Pyle, Robert Michael. *Chasing Monarchs: Migrating with Butterflies of Passage.* Boston: Houghton Mifflin, 1999.

Rankin, Mary, ed. *Migration: Mechanisms and Adaptive Significance.* Port Aransas, Tex.: Marine Science Institute, University of Texas at Austin, 1985.

Reader's Digest Association. *The Wildlife Year.* Pleasantville, N.Y.: Author, 1993.

Schone, Hermann. *Spatial Orientation: The Spatial Control of Behavior in Animals and Man.* Translated by Camilla Strausfeld. Princeton, N.J.: Princeton University Press, 1984.

MIMICRY

Types of ecology: Behavioral ecology; Physiological ecology

Mimicry is the process whereby one organism resembles another and, because of this resemblance, obtains an evolutionary advantage.

The broadest description of mimicry is when one organism, called the operator or dupe, cannot distinguish a second organism, referred to as the mimic, from a third organism or a part of the environment, called the model. There are many different types of mimicry. Some mimics look like another organism; some smell like another organism; some may even feel like another organism. There are also many ways that mimicking another organism could be helpful. Mimicry may help to hide an organism in plain sight or protect a harmless organism from predation when it mimics a harmful organism. It can even help predators sneak up on prey species when the predator mimics a harmless organism.

Camouflage vs. Mimicry

In the case of hiding in plain sight, the line between camouflage and mimicry is not sharply defined. Spots or stripes that help an organism blend with the surroundings is classified as camouflage, because those patterns allow the organism to remain hidden in many areas that have mixtures of sunlight and shadow, and the organism does not look like any particular model. As an organism's appearance begins to mimic another organism more and more closely, rather than displaying just a general pattern, it moves toward mimicry. As in all other areas of biology, there are arguments about where camouflage ends and mimicry begins. The stripes of a tiger and the spots on a fawn are certainly camouflage. The appearance of a stick insect is more ambiguous. Its body is very thin and elongated and is colored in shades of brown and gray. Is this mimicry of a twig or just very good camouflage? Many biologists disagree. The shapes and colors of many tropical insects, especially mantises, also fall into this gray area of either extremely good camouflage or simple mimicry.

Coloration: Batesian and Müllerian Mimicry

In contrast to camouflage, which hides its bearers, many species of dangerous or unpalatable animals are brightly colored. This type of color pattern, which stands out against the background, is called warning coloration. Some examples are the black and white stripes of the skunk, the yellow and black stripes of bees and wasps, red, black, and yellow stripes of the coral snake, and the bright orange of the monarch butterfly. Several species of harmless insects have the same yellow and black pattern that is seen on wasps. In addition to mimicking the coloration of the more dangerous insects, some harmless flies even mimic the wasps' flying patterns or their buzzing sound. In each case, animals that have been stung by wasps or bees avoid both the stinging insects and their mimics. This mimicry of warning coloration is called Batesian mimicry. Batesian mimicry is also seen in the mimicry of the bright red color of the unpalatable red eft stage of newts by palatable salamanders.

Sometimes two or more dangerous or unpalatable organisms look very much alike. In this case, both are acting as models and as mimics. This mimicry is called Müllerian mimicry. Müllerian mimicry is seen in monarch and viceroy butterflies. Both butterflies have in their bodies many of the chemicals found in the plants they ate as larvae. These include many unpalatable chemicals and even toxic chemicals that cause birds to vomit. If a bird eats either a monarch or a viceroy that has these chemicals, the bird usually remembers and avoids preying on either species again—a classic

Monarch butterflies and viceroy butterflies both have in their bodies unpalatable and even toxic chemicals that cause birds to vomit and avoid preying on them later—a classic example of Müllerian mimicry, in which two or more dangerous or unpalatable organisms look very much alike. It is believed that evolutionary processes select for such mimicry because it simplifies the process of recognition and avoidance on the part of predators, thereby increasing the fitness of the similar-looking prey. (PhotoDisc)

Müllerian mimicry. Interestingly, not all monarchs or viceroys are unpalatable. It depends on the types and concentrations of chemicals in the particular plants on which they fed as larvae. Birds that have eaten the palatable monarchs or viceroys do not reject either monarchs or viceroys when offered them as food, but birds that have eaten an unpalatable monarch or an unpalatable viceroy avoid both palatable and unpalatable members of both species. This represents both Batesian and Müllerian mimicry at work.

Aggressive Mimicry

Mimicry by predators is called aggressive mimicry. The reef fish, called the sea swallow, is a cleaner fish, and larger fish enter the sea swallow's territory to be cleaned of parasites. The saber-toothed blenny mimics the cleaner in both appearance and precleaning behavior, but when fish come to be cleaned, the blenny instead bites off a piece of their flesh to eat. Anglerfish have small extensions on their heads that resemble worms. They mimic worms to lure their prey close enough to be eaten. The alligator snapping turtle's tongue and the tips of the tails of moccasins, copperheads, and other pit vipers are also wormlike and are used as lures. Certain predatory female fireflies respond to the light flashes of males of a different species with the appropriate response of the female of that species. This lures the male closer, and when the unsuspecting male is close enough to mate, the female devours him. This mimicry is quite complex, because the predatory females are able to mimic the response signals of several different species.

Octopuses

There are many other instances of mimicry, but the world champion mimics may be octopuses. As predators, these animals show unbelievable aggressive mimicry of other reef organisms. Octopuses can take on the color, shape, and even texture of corals, algae, and other colonial reef dwellers. As a prey species, the octopus can use the same type of mimicry for camouflage, but can also be a Batesian mimic, taking on the color and shape of many of the reef's venomous denizens.

Since in each case, being a mimic helped the organism in some way, it is not hard to understand how mimicry may have evolved. In a population in which some organisms were protected by being mimics, the protected mimics were most likely to mate and leave their genes for the next generation while the unprotected organisms were less likely to breed.

Richard W. Cheney, Jr.

See also: Altruism; Camouflage; Communication; Defense mechanisms; Displays; Mammalian social systems; Pheromones; Poisonous animals; Predation; Reproductive strategies; Territoriality and aggression.

Sources for Further Study
Brower, Lincoln P., ed. *Mimicry and the Evolutionary Process.* Chicago: University of Chicago Press, 1988.
Ferrari, Marco. *Colors for Survival: Mimicry and Camouflage in Nature.* Charlottesville, Va.: Thomasson-Grant, 1993.
Owen, D. *Camouflage and Mimicry.* Chicago: University of Chicago Press, 1982.
Salvato, M. "Most Spectacular Batesian Mimicry." In *University of Florida Book of Insect Records*, edited by T. J. Walker. Gainesville: University of Florida Department of Entomology and Nematology, 1999.
Wickler, Wolfgang. *Mimicry in Plants and Animals.* New York: McGraw-Hill, 1968.

MOUNTAIN ECOSYSTEMS

Types of ecology: Biomes; Ecosystem ecology

Mountains cover one fifth of the earth's terrestrial surface, and they are one of the most extreme environments in the global ecosystem. Mountains are globally significant landforms that function as storehouses for irreplaceable resources such as clean air and water, biological and cultural diversity, as well as timber and mineral resources.

Mountains are the most conspicuous landforms on earth. They are found on every continent and have been defined simply as elevated landforms of high local relief, with much of the surface in steep slopes, displaying distinct variations in climate and vegetation zones from the base to the summit. The earth's mountain ranges have been created by the collision of tectonic plates. Associated with many of these mountain ranges are volcanoes. If the solidified magma of a volcano builds up, it can become a mountain; likewise, if the collision involves two oceanic plates, a string of volcanic mountains, called an island arc, can form on the ocean floor.

Mountain Habitat
Mountains are globally significant reservoirs of biodiversity. They contain rich assemblages of species and ecosystems. Because of the rapid changes in altitude and temperature along a mountain slope, multiple ecological zones are stacked upon one another, sometimes ranging from dense tropical jungles to glacial ice within a few kilometers. Many plant and animal species are found only on mountains, having evolved over centuries of isolation to inhabit these specialized environments. Mountains can also function as biological corridors, connecting isolated habitats or protected areas and allowing species to migrate between them. These extraordinary ecological conditions, coupled with many bio-climatic zones, have resulted in a high number of ecological niches available for habitation in mountain ecosystems.

Biodiversity
Because of the great diversity in habitats within mountainous regions, with each region showing a different combination of environmental factors, total mountain fauna is relatively rich and the variety of small communities very great, in spite of the general severity of the mountain environment as a whole. Likewise, this diversity has resulted in a wide range of

Mountains are perhaps the most significant reservoirs of biodiversity. Because of the rapid changes in altitude and temperature along a mountain slope, multiple ecological zones are stacked upon one another—sometimes ranging from dense tropical jungles to glacial ice—and therefore contain rich assemblages of species and ecosystems. (PhotoDisc)

endemic species that have evolved over centuries of isolation from other genetic material. Rocky Mountain National Park typifies this diversity as a home to some 900 species of plants, 250 species of birds, and 60 species of mammals. Some are easily seen and others are elusive, but all are part of the ecosystem in the park. On a global scale, the diversity of mountain fauna extends to many species of ungulates, including elk, bighorn sheep, moose, and deer. Also present in mountain communities are many species of rodents. Rodent species may include beaver, marmots, squirrels, and chipmunks. Other mammalian animal life include bear, canids, such as coyote and wolf, and many species of felids, such as mountain lions and bobcats. Mountain avian fauna comprise many families of hummingbirds, bluebirds, hawks, falcons, eagles, and many more.

Ecological Threats
Mountains are threatened in a variety of ways, but without question, human settlement and activities such as camping, hiking, and other recreational activities constitute the biggest threats to the mountain ecosystem.

Hikers and motorized offroad vehicles, for example, create tracks in the soil that form erosion gullies and trample vegetation that has taken many years to grow. Commercial harvesting of trees in the lower forest zones of mountains is having an increasingly detrimental effect on biodiversity. Global warming is another threat to mountain ecosystems. Snowlines are receding, and continued melting of glaciers and polar ice caps could eventually lead to drying of major river systems which feed from them.

In an attempt to restore or conserve mountain ecosystems, many countries have replanted indigenous trees with fast-growing coniferous trees, in an ill-fated effort to supply a growing human population with wood products. These hybrid forests are not nearly as beautiful as the native forests, but more to the point, they do not offer environments conducive to the ecosystems that the native species supported. This loss of habitat creates a loss of wildlife, which then becomes threatened and eventually endangered because of the decline of native vegetation.

Jason A. Hubbart

See also: Biomes: types; Forest fires; Forest management; Forests; Grazing and overgrazing; Habitats and biomes; Lakes and limnology; Old-growth forests; Rain forests; Rain forests and the atmosphere; Restoration ecology; Savannas and deciduous tropical forests; Sustainable development; Taiga; Tundra and high-altitude biomes; Wildlife management.

Sources for Further Study
Denniston, D. *High Priorities: Conserving Mountain Ecosystems and Cultures.* Worldwatch Paper 123. Washington, D.C.: Worldwatch Institute, 1995.

Messerli, B., and J. D. Ives. *Mountains of the World: A Global Priority.* New York: Parthenon, 1997.

Price, L. *Mountains and Man.* Berkeley: University of California Press, 1981.

Sauvain, P. *Geography Detective: Mountains.* Minneapolis: Carolrhoda Books, 1996.

Stronach, N. *Mountains.* Minneapolis: Lerner, 1995.

MULTIPLE-USE APPROACH

Types of ecology: Agricultural ecology; Restoration and conservation ecology

The multiple-use approach is a concept of resource use in which land supports several concurrent managed uses rather than single uses over time and space.

The multiple-use approach is a management practice that is teamed with the concept of sustained yield. Multiple use began as a working policy, generally associated with forestry, and was enacted as law in 1960. As a concept of land-use management, it has most often been applied to the use of forestlands. Historically, multiple use has been linked with another concept, that of sustained yield.

Historical Background
The history of the intertwined multiple-use and sustained-yield approaches to land management in the United States dates from the late 1800's. Prior to that time, forestlands were used for timber production, rangeland for grazing, and parklands for recreation. Little attention was given to the interrelated aspects of land use. By the late 1800's, however, some resource managers began to see land as a resource to be managed in a more complex, integrated fashion that would lead to multiple use. This awakening grew out of the need for conservation and sustained yield, especially in the forest sector of the resource economy.

Sustained Yield
Since the earliest European settlement of North America, forest resources had been seen both as a nearly inexhaustible source of timber and as an impediment to be cleared to make way for agriculture. This policy of removal led to serious concern by the late 1800's about the future of American forests. By 1891 power had been granted to U.S. president Benjamin Harrison to set aside protected forest areas. Both he and President Grover Cleveland took action to establish forest reserves. To direct the management of these reserves, Gifford Pinchot was appointed chief forester. Pinchot was trained in European methods of forestry and managed resources, as noted by Stewart Udall in *The Quiet Crisis* (1963), "on a sustained yield basis." The sustained-yield basis for forest management was thus established. Essentially, the sustained-yield philosophy holds that the amount of timber harvested should not exceed the ultimate timber growth during the same period.

Multiple Use

Properly managed, forestlands can meet needs for timber on an ongoing, renewable basis. However, land in forest cover is more than a source of timber. Watersheds in such areas can be protected from excessive runoff and sedimentation through appropriate management. Forest areas are also wildlife habitat and potential areas of outdoor recreation. The combination of forest management for renewable resource production and complex, interrelated land uses provided the basis for the development of multiple-use sustained-yield as a long-term forest management strategy.

Multiple Use Joins Sustained Yield

The merging of these two concepts took shape over a period of many years, beginning in the early twentieth century. The establishment of national forests by Presidents Harrison and Cleveland provided a base for their expansion under President Theodore Roosevelt in the early 1900's. With the active management of Pinchot and the enthusiastic support of Roosevelt, the national forests began to be managed on a long-term, multiple-use sustained-yield basis. The desirability of this approach eventually led to its formalization by law: On June 12, 1960, Congress passed the Multiple Use-Sustained Yield Act. To some, this act was the legal embodiment of practices already in force. However, the act provides a clear statement of congressional policy and relates it to the original act of 1897 that had established the national forests.

The 1960 act specifies that "the national forests are established and shall be administered for outdoor recreation, range, timber, watershed, and wildlife and fish purposes." Section 2 of the act states that the "Secretary of Agriculture is authorized and directed to develop and administer the renewable resources of the national forests for multiple use and sustained yield of the several products and services obtained therefrom." The act gives no specifics, providing a great deal of freedom in choosing ways to meet its provisions. It also refrains from providing guidelines for management. In practice, the achievement of a high level of land management under the act has called for advocating a conservation ethic, soliciting citizen participation, providing technical and financial assistance to public and private forest owners, developing international exchanges on these management principles, and extending management knowledge.

Jerry E. Green

See also: Biopesticides; Conservation biology; Erosion and erosion control; Genetically modified foods; Grazing and overgrazing; Integrated pest management; Old-growth forests; Rangeland; Reforestation; Restoration ecology; Slash-and-burn agriculture; Soil; Soil contamination; Sustainable development.

Sources for Further Study

Cutter, Susan, Hilary Renwick, and William Renwick. *Exploitation, Conservation, Preservation: A Geographical Perspective on Natural Resource Use.* 3d ed. New York: J. Wiley & Sons, 1999.

Hewett, Charles E., and Thomas E. Hamilton, eds. *Forests in Demand: Conflicts and Solutions.* Boston: Auburn House, 1982.

Lovett, Francis. *National Parks: Rights and the Common Good.* Lanham, Md.: Rowman & Littlefield, 1998.

Sedjo, Roger A., ed. *A Vision for the U.S. Forest Service: Goals for Its Next Century.* Washington, D.C.: Resources for the Future, 2000.

Udall, Stewart. *The Quiet Crisis.* 1963. Reprint. Salt Lake City, Utah: Gibbs Smith, 1998.

MYCORRHIZAE

Type of ecology: Community ecology

Mycorrhizae are mutualistic, symbiotic relationships between plant roots or other underground organs and fungi. They are among the most abundant symbioses in the world.

Mycorrhizal associations (from the Greek *mukes*, meaning "fungus," and *rhiza*, meaning "root") have been described in virtually all economically important plant groups. Investigators in Europe detected fungal associations in most European species of flowering plants, all gymnosperms, ferns, and some bryophytes, especially the liverworts. Similar patterns are predicted in other ecosystems. Continuing studies of ecosystems, from boreal forests to temperate grasslands to tropical rain forests and agroecosystems, also suggest that most plant groups are intimately linked to one or more species of fungus.

It is theorized that most of the plants in stable habitats where competition for resources is common probably have some form of mycorrhizal association. Species from all the major taxonomic groups of fungi, including the *Ascomycotina*, *Basidiomycotina*, *Deuteromycotina*, and *Zygomycotina*, have been found as partners with plants in mycorrhizae.

Considering the prevalence of mycorrhizae in the world today, botanists theorize that mycorrhizae probably arose early in the development of land plants. Some suggest that mycorrhizae may have been an important factor in the colonization of land. The fungal partner (or mycobiont) in a mycorrhizal relationship benefits by gaining a source of carbon. Often these mycobionts are poor competitors in the soil environment. Some mycobionts have apparently coevolved to the point that they can no longer live independently of a plant host.

The plant partner in the mycorrhizal relationship benefits from improved nutrient absorption. This may occur in different ways; for example, the mycobiont may directly transfer nutrients to the root. Infected roots experience more branching, thus increasing the volume of soil that the plant can penetrate and exploit. Evidence also suggests that mycorrhizal roots may live longer than roots without these associations. Comparison of the growth of plants without mycorrhizae to those with fungal partners suggests that mycorrhizae enhance overall plant growth.

Endomycorrhizae

Mycorrhizae may be classified into two broad groups: endomycorrhizae and ectomycorrhizae. Endomycorrhizae enter the cells of the root cortex. Ectomycorrhizae colonize plant roots but do not invade root cortex cells.

The most common form of endomycorrhizae are the vesicular-arbuscular mycorrhizae. The fungi involved are zygomycetes. These mycorrhizae have internal structures called arbuscules, which are highly branched, thin-walled tubules inside the root cortex cells near the vascular cylinder. It is estimated that 80 percent of all plant species may have vesicular-arbuscular mycorrhizae. This type of mycorrhiza is especially important in tropical trees.

There are several other subtypes of endomycorrhizae. Ericoid mycorrhizae, found in the family *Ericaceae* and closely related families, supply the host plants with nitrogen. These are usually restricted to nutrient-poor, highly acidic conditions, such as heath lands. Arbutoid mycorrhizae, found in members of the *Arbutoideae* and related families, share some similarities with ectomycorrhizae in that they form more developed structures called the sheath and Hartig net (described below).

Monotropoid mycorrhizae, found in the plant family *Monotropaceae*, are associated with plants that lack chlorophyll. The host plant is completely dependent on the mycobiont, which also has connections to the roots of a nearby tree. Thus the host, such as *Monotropa*, indirectly parasitizes another plant by using the mycobiont as an intermediate. Orchidaceous mycorrhizae are essential for orchid seed germination.

Ectomycorrhizae

Ectomycorrhizae are common in forest trees and shrubs in the temperate and subarctic zones. Well-developed fungal sheaths characterize these mycorrhizae, along with special structures called Hartig nets. Basidiomycetes are the usual mycobionts and often form mushrooms or truffles. Ectomycorrhizae help protect the host plant from diseases by forming a physical fungal barrier to infection.

The Fungal Partner

Individual filaments of a fungal body are called hyphae. The entire fungal body is called a mycelium. Root infection may occur from fungal spores that germinate in the soil or from fungal hyphae growing from the body of a nearby mycorrhiza. When infection occurs, hyphae are drawn toward certain chemical secretions from a plant root.

In ectomycorrhizae, root hairs do not develop in roots after infection occurs. Infected roots have a fungal sheath, or mantle, that ranges from 20 to

40 micrometers thick. Fungal hyphae penetrate the root by entering between epidermal cells. These hyphae push cells of the outer root cortex apart and continue to grow outside the cells. This association of hyphal cells and root cortex cells is called a Hartig net. In ectomycorrhizae, the mycobionts never invade plant cells, nor do they penetrate the endodermis or enter the vascular cylinder. The root tip may be ensheathed by fungi, but the apical meristem is never invaded. Main roots experience fewer anatomical changes than lateral roots after infection. Lateral roots become thickened, may show the development of characteristic pigments, and grow very slowly. Infected roots also show different branching patterns, compared to those of uninfected roots.

Endomycorrhizae are highly variable in structure. Many endomycorrhizae do not have sheaths or Hartig nets. In all endomycorrhizae, hyphae penetrate into root cortex cells, while portions of the mycelium remain in contact with the soil. The hyphae that remain in the soil are important in fungal reproduction and produce large numbers of haploid spores. Fungi do not invade root meristems, vascular cylinders, or chloroplast-containing cells in the plant.

Some of the host cells contain fungal extensions called vesicles that are filled with lipids. Vesicles are specialized structures that are often thick-walled and may serve as storage sites or possibly in reproduction. Vesicles are also produced on the hyphae that grow in the soil. Near the vascular cylinder, the hyphae branch dichotomously and form large numbers of thin-walled tubules called arbuscules that invade host cells. The arbuscules cause the host membranes to fold inward, creating a plant-fungus interface that has a very large surface area. The arbuscules last for about fourteen days before they break down on their own or are digested by the host cell. Host cells whose fungal arbuscules have broken down may be reinvaded by other hyphae.

Darrell L. Ray

See also: Coevolution; Communities: ecosystem interactions; Communities: structure; Lichens; Old-growth forests; Symbiosis.

Sources for Further Study

Deacon, J. W. *Modern Mycology.* 3d ed. Malden, Mass.: Blackwell Science, 1997.
Harley, J. L., and S. E. Smith. *Mycorrhizal Symbiosis.* New York: Academic Press, 1983.
Raven, Peter H., Ray F. Evert, and Susan E. Eichhorn. *Biology of Plants.* 6th ed. New York: W. H. Freeman/Worth, 1999.

NATURAL SELECTION

Types of ecology: Evolutionary ecology; Speciation

Natural selection is the process of differential survival and reproduction of individuals resulting in long-term changes in the characteristics of species. This process is central to evolution.

Natural selection is a three-part process. First, there must exist differences among individuals in some trait. Second, the trait differences must lead to differences in survival and reproduction. Third, the trait differences must have a genetic basis. Natural selection results in long-term changes in the characteristics of a population. As one of the central processes responsible for evolution, natural selection results in both fine-tuning adaptations of populations and species to their environments and creating differences among species.

The importance of natural selection was first recognized by Charles Darwin as the primary mechanism for evolutionary change. Processes that support natural selection include genetic drift and migration. These processes interact with the processes responsible for producing variation (mutation and development) and those responsible for determining the rate and direction of evolution (mating system, population size, and long-term ecological changes) to establish the evolutionary path of a species.

The Basic Process

Natural selection occurs through the interaction of three conditions: variation among individuals in a population in some trait, differences in those individuals' fitness as a result of the variations in that trait, and heritable variation in that trait. If those three conditions are met, then the characteristics of the population with respect to that trait will change from one generation to the next until equilibrium with other processes is reached.

An example that demonstrates this process involves the peppered moth. It has two forms in the United Kingdom, a light-colored form and a dark-colored form; there is variation in color among individuals. Genetic analysis has shown that this difference in color is caused by a single gene; the variation has a heritable basis. The moth is eaten by birds that find their food by sight. The light-colored form cannot be seen when sitting on lichen-covered trees, while the dark-colored form can be seen easily. Air pollution kills the lichen, however, and turns the trees dark in color. Then, the dark-colored form is hidden and the light-colored form visible. Thus, differences in color lead to fitness differences: Dark moths will become more fit if air pollution increases, renders them less visible to predators, and thereby makes it likelier that their genetic material will pass into another generation. In fact, this proved to be the case: In the early nineteenth century, the dark-colored form was very rare. In the last half of the nineteenth century, however, air pollution increased, and the dark-colored form became much more frequent as a result of natural selection.

Directional Selection

The characteristics of a population can be changed by natural selection in several ways. If individuals in a population with an extreme value for a trait have the greatest fitness on average, then the mean value of the trait will change in a consistent direction, which is called directional selection. For example, the soil in the vicinity of mines contains heavy metals that are toxic to plants. Individuals with the greatest resistance to heavy metals have the highest survivorship. Evolution leads to an increase in resistance.

Stabilizing Selection

If individuals in a population with intermediate values for a trait have the greatest fitness on average, then the variation in the trait will be reduced, which is called stabilizing selection. For example, in many species of birds, individuals with intermediate numbers of offspring have the greatest fitness. If an individual has a small number of offspring, that parent has reduced reproduction and a low fitness. If the number of offspring is large, the parent will not be able to provide enough food for all the young, and most, or all, will starve, again resulting in reduced reproduction and a low fitness. Evolution leads to all birds producing the same, intermediate number of offspring.

Disruptive Selection

If individuals in a population with different values for a trait have the greatest fitnesses on average and intermediates have low fitness, then the variation in the trait will be increased. This is called disruptive selection. For example, for Darwin's finches, individuals with long, thin bills are able to probe into rotting cacti to find insects. Individuals with short, thick bills are able to crack hard seeds. Individuals with intermediate-shaped bills are not able to do either well and have reduced fitness relative to the more extreme types. Evolution therefore leads to two different species of finch with different bills.

Natural selection is a slow process. The rate of evolution—that is, response to selection—is determined by the magnitude of fitness differences among individuals and the heritability of traits. Fitness differences tend to be small so that more fit individuals on average may have only a few more offspring than less fit individuals. Heritabilities of most traits are low to intermediate, meaning that most differences among individuals are not a result of genetic differences. Therefore, even if one individual has many offspring and another has few offspring, they may not differ genetically and no change will occur. For example, if all the beetles in a population were between one and two centimeters in length and there was selection for larger beetles, it could take five hundred generations before all beetles were larger than two centimeters. Also, the direction of selection may change from one generation to the next, so that no net change occurs.

Correlational Selection

Natural selection does not act on traits in isolation. How a trait affects fitness in combination with other traits—called correlational selection—is important. For example, fruit flies lay their eggs in rotting fruit. Considered in isolation, a female should always lay as many eggs as possible. One fruit is not big enough for all the eggs she might lay, however, so she must fly from fruit to fruit. Flying requires energy, and the more energy that is used in flight the less that can be used to make eggs. Hence, natural selection results in the division of energy between eggs and flight that yields the greatest overall number of offspring. This example demonstrates that the result of natural selection is often a trade-off among different traits.

Sexual Selection

By acting differently on males and females, natural selection results in sexual selection. This form of selection can explain differences in the forms of males and females of a species. In general, because male gametes, sperm, are much smaller and "cheaper" to produce than female gametes, eggs, more sperm than eggs are produced. As a result, it is possible for one male to fertilize many eggs, while other males fertilize few or no eggs. For example, a lion pride usually consists of one or a few males and many females. Other males are excluded, and they live separately; larger males are able to chase away smaller males. The thick mane on male lions helps to protect their throats when they fight other males. Thus, larger males with thicker manes tend to survive, fathering more cubs than other males, leading to additional bias in selection favoring these traits. This is an example of sexual selection because the only selection pressure for the trait in question is on males; all females, regardless of size, will mate. The result is that males are larger than females and have manes.

Group and Kin Selection

Natural selection can occur not only among individuals but also among groups. This process is generally known as group selection; when the groups are composed of related individuals, it is called kin selection. Group selection operates the same way as individual selection. The same three conditions are necessary: variation among groups in some trait, fitness differences among groups because of that trait, and a heritable basis for that trait.

For example, in Australia, rabbits introduced from Europe in 1859 spread rapidly during the next sixty years. In order to control the rabbits, a virus was introduced in 1950. At first, the virus was very virulent, killing almost all infected animals within a few days. After ten years, however, the virus had evolved to become more benign, with infected rabbits living longer or not becoming sick at all. Virulent strains of the virus grow and reproduce faster than benign strains. Therefore, within a single rabbit, virulent strains have a higher fitness than benign strains. The longer a rabbit lives, however, the more opportunity there is for the virus to be passed to other rabbits. Thus, a group of benign viruses infecting a rabbit are more likely to be passed on than a group of virulent viruses. In this example, group selection among rabbits resulted in evolution opposite to individual selection within rabbits; however, group selection and individual selection can result in evolution in the same direction. In general, natural selection can act at many levels: the gene, the chromosome, the individual, a group of individuals, the population, or the species.

Measuring Natural Selection

Natural selection is investigated in two ways: through indirect measurement and through direct measurement. Indirect methods involve observing the outcome of natural selection and inferring its presence. Direct methods involve measuring the three parts of the process and following the course of evolution. Although the direct methods are preferred, as they provide direct proof of natural selection, in most instances, only indirect methods can be used.

Indirect methods involve three kinds of observations. First, comparisons are made of trait similarities or differences among populations or species living in the same or different areas. For example, many species of animals living in colder climates have larger bodies than those living in warmer climates. It is inferred, therefore, that colder climates result in natural selection for larger bodies. Second, long-term studies are done of traits, in particular changes in a group in the fossil record. For example, during the evolution of horses, their food, grasses, became tougher and horses' teeth became thicker. It is inferred, therefore, that tough grass resulted in natural selection for thicker teeth. Third, comparisons are made of gene frequencies of natural populations, with predictions from mathematical models. Gene frequencies are measured using various techniques, including scoring differences in appearance, as with light-colored and dark-colored moths; using electrophoresis to observe differences in proteins; and determining the sequence of base pairs of deoxyribonucleic acid (DNA). The models make predictions about expected frequencies in the presence or absence of selection. Indirect methods are best at revealing long-term responses to evolution and general processes of natural selection that affect many species. The indirect methods suffer from the problem that often many processes will result in similar patterns. So, it must be assumed that other processes were not operating, or other predictions must be made to separate the processes.

Direct methods involve two kinds of observation. First, there is observation of changes in a population following some change in the environment. There are many types of environmental changes, including human-made changes, natural disasters, seasonal changes, and introductions of species into new environments. For example, from the changes in the peppered moth following a change in pollution levels, one can measure the effects of natural selection. The second type of observation is the direct measurement of fitness differences among individuals with trait differences. For example, individual animals are tagged at an early age and survival and reproduction are monitored. Then, statistical techniques are used to find a relationship between fitness and variation among individuals in some trait. Alternatively, comparisons of traits are made between groups of individuals, such as breeding and nonbreeding, adults and juveniles, or live and dead individuals, again using statistical techniques. For example, lions that breed are larger than lions that do not breed. Direct methods are best at revealing the relative importance to natural selection of the three factors (variation, fitness differences, and heritability). The direct methods suffer from two limitations. It takes a long time for evolution to occur. So, although one can measure natural selection, it is often not known if it results in evolution. Also, for many species, it is impossible or impractical to mark individuals and follow them through their lives.

Many methods can be used to study natural selection and evolution. Each method provides information about different parts of the process. Only through the integration of these methods can the entire process of evolution be revealed.

Knowledge of natural selection is still growing; many questions proposed by Darwin and others are yet to be answered. It is still not known to what extent organisms are well adapted to their environments or whether the evolution of the parts of the chromosome that are not translated into proteins are a result of processes that do not involve natural selection. Of the many theories of how natural selection works, it is still unknown which ones are the most important in nature and to what extent evolution is caused by natural selection at the level of the individual, the group, and the species.

Ecological Implications

Scientists and researchers combined their knowledge of natural selection with a revolutionary breakthrough in molecular biology in the 1950's: the discovery of the helical structure of deoxyribonucleic acid (DNA) and subsequent discoveries and developments surrounding recombinant DNA technology. The resultant manipulation of traits in organisms from crops to mammals led to an age of genetic engineering. Today, strawberries, soybeans, dairy products, and a host of other foods, as well as higher organisms including human tissues, can be genetically manipulated to beneficial, as well as often unintended negative, ends. The addition of a new gene into an organism will result in natural selection on that gene and change selection on other genes. For example, certain crop plants genetically modified to resist herbicides have begun to mix with native species considered weeds, creating a new problem of "superweeds" resistant to herbicides.

An understanding of natural selection is also critical for conservation biology. During the twentieth century, the rate at which natural areas are being destroyed and species are becoming extinct has accelerated tremendously. Conservation biology attempts to stop that destruction and preserve species diversity. For extinction of endangered species to be halted, it must be understood how natural selection will affect these species given massive environmental changes. By discovering how evolution is occurring under natural conditions, researchers are learning how to design nature preserves to maintain species.

Samuel M. Scheiner, updated by Christina J. Moose

See also: Adaptations and their mechanisms; Adaptive radiation; Coevolution; Colonization of the land; Convergence and divergence; Dendrochronology; Development and ecological strategies; Evolution: definition and theories; Evolution: history; Evolution of plants and climates; Extinctions and evolutionary explosions; Gene flow; Genetic drift; Genetically modified foods; Isolating mechanisms; Nonrandom mating, genetic drift, and mutation; Paleoecology; Population genetics; Punctuated equilibrium vs. gradualism; Speciation; Species loss.

Sources for Further Study

Avers, Charlotte J. *Process and Pattern in Evolution*. New York: Oxford University Press, 1989.

Bell, Graham. *Selection: The Mechanism of Evolution*. New York: Chapman and Hall, 1997.

Brandon, Robert N. *Adaptation and Environment*. Princeton, N.J.: Princeton University Press, 1990.

Darwin, Charles. *On the Origin of Species by Means of Natural Selection*. London: J. Murray, 1859.

Endler, John A. *Natural Selection in the Wild*. Princeton, N.J.: Princeton University Press, 1986.

Futuyma, Douglas J. *Evolutionary Biology*. 3d ed. Sunderland, Mass.: Sinauer Associates, 1998.

Gould, Stephen J. *The Panda's Thumb*. New York: W. W. Norton, 1980.

Pianka, Eric R. *Evolutionary Ecology*. Boston: Addison-Wesley, 1999.

Provine, William B. *Sewall Wright and Evolutionary Biology*. Chicago: University of Chicago Press, 1986.

NONRANDOM MATING, GENETIC DRIFT, AND MUTATION

Types of ecology: Evolutionary ecology; Population ecology

Nonrandom mating, genetic drift, and mutation are three mechanisms, besides natural selection and migration, that can change the genetic structure of a population.

Evolution is a process in which the gene frequencies of a population change over time, and nonrandom mating, genetic drift, and mutation are all mechanisms of genetic change in populations. These mechanisms violate the assumptions of the Hardy-Weinberg model of genetic equilibrium by increasing or decreasing the frequency of heterozygote genotypes in the population.

Nonrandom Mating

Nonrandom mating occurs in a population whenever every individual does not have an equal chance of mating with any other member of the population. While many organisms do tend to mate randomly, there are some common patterns of nonrandom mating. Often, individuals tend to mate with others nearby, or they may choose mates that are most like themselves. When individuals choose mates that are phenotypically similar, positive assortative mating has occurred. If mates look physically different, then it is negative assortative mating. Population geneticists use the term "assortative" because it means "to separate into groups," usually in a pattern that is not random. The terms "positive" and "negative" refer to the probability that mated pairs have the same phenotype more or less often than expected by chance. Two color varieties of snow geese (*Chen hyperborea*), blue and white, are commonly found breeding in Canada, and they show positive assortative mating patterns based on color. The geese tend to mate only with birds of the same color; blue mate with blue and white with white. Since a bird's color (phenotype) is determined by the presence of a dominant blue color allele, matings between similar phenotypes are also matings between similar genotypes. Matings between similar genotypes cause the frequency of individuals that are homozygous for the blue or the white allele to be greater, and the frequency of heterozygotes to be less than if mating were random and in Hardy-Weinberg equilibrium. Negative assortative mating increases the frequency of heterozygote genotypes in the population and decreases homozygote frequency.

Assortative mating does not change the frequency of the blue or white alleles in the goose population; it simply reorganizes the genetic variation and shifts the frequency of heterozygotes away from Hardy-Weinberg equilibrium frequencies.

Inbreeding

Inbreeding is the mating of relatives and is similar to positive assortative mating because like genotypes mate and result in a high frequency of homozygotes in the population. In assortative mating, only those genes that influence mate choice become homozygous, but inbreeding increases the homozygosity of all the genes. High homozygosity means that many of the recessive alleles that were masked by the dominant allele in heterozygotes will be expressed in the phenotype. Deleterious or harmful alleles can remain hidden from selection in the heterozygote, but after one generation of inbreeding, these deleterious alleles are expressed in a homozygous condition and can substantially reduce viability below normal levels. Low viability resulting from mating of like genotypes is called inbreeding depression.

Genetic Drift

Genetic drift, like positive assortative mating, reduces the frequency of heterozygotes in a population, but with genetic drift, the frequency of alleles in a population changes. Nonrandom mating does not change allele frequency. Genetic drift is sometimes called random genetic drift because the mechanism of genetic change is random and attributable to chance events in small populations, such that allele frequencies tend to wander or drift. Statisticians use the term "sampling effect" to describe observed fluctuations from expected values when only a few samples are chosen, and it is easy to observe by tossing a coin. A fair coin flipped a hundred times would be expected to produce approximately fifty heads and fifty tails, plus or minus a few heads or tails. Yet, if the coin is flipped only four times, it is not too surprising to get four heads or four tails. The probability of getting either all heads or all tails on four consecutive flips is one out of eight, but the probability of getting all heads or all tails decreases to much less than one in a billion as the sample size increases from four to a hundred tosses. Similarly, it is much easier for nonrandom events to occur in small populations than in large populations. If a population has two alleles with equal frequency for a particular trait, then the result of random mating can be simulated by tossing a coin. The frequency of each allele in the next generation would be determined by flipping the coin twice for each individual, since sexually reproducing organisms have two alleles for each trait, and counting the number of heads and tails. In a small population, only a few gametes, each containing one allele for the trait, will fuse to form zygotes. Chance events can cause the frequencies of alleles in a small population to drift randomly from generation to generation; often one allele is lost from the population.

In small populations with fewer than fifty mating pairs, alleles may be eliminated in fewer than twenty generations by random genetic drift, leaving only one allele for a particular trait in the population. Thus, all individuals would be homozygous for the remaining allele and genetically identical. Theoretically, in any finite population random genetic drift will occur, but it is usually negligible if the population size is greater than a hundred. Sometimes, disasters or disease may drastically reduce the population size, causing a bottleneck effect. The bottleneck in population size reduces genetic variability in a population because there are only a few alleles and results in random genetic drift. Many islands and new populations are established by a small group of founders that constitute a nonrandom genetic sample because they have only a fraction of the alleles from the original large population. Founder effects and bottleneck effects are phenomena that result in a loss of heterozygosity and decreased genetic variability because of the chance drift in allele frequency away from Hardy-Weinberg equilibrium values in small populations.

Mutations

Mutations are any changes in the genetic material that can be passed on to offspring. Some mutations are changes at a single point in the chromosome, while at other times, pieces of genetic material are removed, extra pieces are added, or pieces are exchanged with other chromosomes. All these changes could result in the formation of new alleles or could change one allele into a different allele. The random mistakes in the chromosomes occur at the molecular level, and only later are the changes in information or alleles translated into phenotypic differences. Thus, mutation is the ultimate source of genetic variability and is random with respect to the needs of the organism. Most mutations are lethal and are never expressed, but nonlethal mutations provide the necessary variation for natural selection. Even though mutations are very important for evolution, they have only a small effect on allele and genotype frequencies in populations because mutation rates are relatively low. If an allele makes up 50 percent of the gene pool and mutates to another allele once for every hundred thousand gametes, it would take two thousand generations to reduce the frequency of the allele by 2 percent. The net effect of mutations is to increase genetic variability, but at a very slow rate.

Studying Genetic Variability

Population geneticists use a wide variety of laboratory, field, and natural experiments to investigate genetic variability. Natural experiments are situations that have developed without a scientist intentionally designing an experiment, but conditions are such that scientists can test a theory. Researchers have used known pedigrees or ancestral histories of zoo animals and have found that mortality rates of inbred young are often two to three times higher than for noninbred young. Population geneticists use pedigrees to calculate the probability that two alleles are identical by descent; this research provides an index of the amount of inbreeding in a population.

The study of random genetic drift is usually carried out in the laboratory. Scientists often use small organisms that reproduce quickly, such as fruit flies (*Drosophila melanogaster*), to conserve space and save time. In a 1956 study of eye color conducted by Peter Buri, after only eighteen generations and sixteen fruit flies per population, more than half of the 107 populations started had only one of the two alleles for eye color.

Mutations are so rare that even fruit flies reproduce too slowly for scientists to study the effects of mutations on populations, even though much is known about the mechanism of mutation by studying *Drosophila*. Small bacterial growth chambers can hold many millions of bacterial cells, and, since they reproduce quickly, even mutations that occur in only one in a million cells can be detected. In 1955, it was found that mutation rates were very low in bacteria until caffeine was added to the growth chamber, whereupon mutation rates increased tenfold. Any chemical or type of radiation that can cause mutations is called a mutagen. Electrophoresis has also been a useful tool for the study of nonrandom mating, genetic drift, and mutations, because allele and genotype frequencies can be determined from samples of the population and unique alleles can be identified.

The Dangers of Inbreeding

Most governments and religions forbid marriages between close relatives because matings between first cousins result in a 20 percent decrease in heterozygosity; for those between brothers and sisters, there is an 80 percent decrease in heterozygosity. The decrease in heterozygosity and genetic variation and increase in homozygote frequency often result in inbreeding depression because deleterious recessive alleles are expressed. All inbreeding is not undesirable; many of the prizewinning bulls and pigs at state fairs have some inbreeding in their pedigrees. Most breeds of dogs were produced by breeding close relatives so that the offspring would have particular traits.

Zookeepers and others that breed and protect rare and endangered species must continually be concerned about the negative effects of both inbreeding and genetic drift. Most zoos are lucky if they have two or three pairs of breeding adults, and total population sizes are usually very small compared to those of natural populations. These conditions mean that inbreeding may reduce the vigor of the population and genetic drift will reduce the diversity of alleles in the population, thus reducing the chances of survival for the captive species. There is hope for rare and endangered species if independent inbred lines are crossed, thus reducing the effects of inbreeding depression, and if breeding adults from other zoos or populations are traded occasionally, thus increasing the effective population size.

Mutations are the ultimate source of genetic variation and so are very important in the study of evolution, but the population-level effects of one mutation are difficult to study because of the low frequency of natural mutations. Certain nonlethal mutations may have little evolutionary impact but may be important medically because spontaneous mutations result in hemophilia or dwarfism (achondroplasia) in more than 3 out of 100,000 cases. As exposure to background radiation and chemical levels increases, mutation rates are likely to increase, as well as the incidence of mutation-related diseases.

William R. Bromer

See also: Biodiversity; Gene flow; Genetic diversity; Genetic drift; Pollination; Population genetics; Speciation; Zoos.

Sources for Further Study

Ayala, Francisco J. *Population and Evolutionary Genetics: A Primer.* Menlo Park, Calif.: Benjamin/Cummings, 1982.

Crow, J. F. *Basic Concepts in Population, Quantitative, and Evolutionary Genetics.* New York: W. H. Freeman, 1986.

Fisher, R. A. *The Genetical Theory of Natural Selection.* New York: Dover, 1958.

Hartl, Daniel L. *A Primer of Population Genetics.* 3d ed. Sunderland, Mass.: Sinauer Associates, 2000.

Mettler, L. E., T. G. Gregg, and H. E. Schaffer. *Population Genetics and Evolution.* 2d ed. Englewood Cliffs, N.J.: Prentice-Hall, 1988.

Real, Leslie A., ed. *Ecological Genetics.* Princeton, N.J.: Princeton University Press, 1994.

Wilson, Edward O., and W. H. Bossert. *A Primer of Population Biology.* Sunderland, Mass.: Sinauer Associates, 1971.

NUTRIENT CYCLES

Types of ecology: Ecoenergetics; Ecosystem ecology

Within an ecosystem, nutrients move through biogeochemical cycles. Those cycles involve chemical exchanges of elements among the earth's atmosphere, water, living organisms, soil, and rocks.

All biogeochemical cycles—whether the carbon cycle, the hydrologic cycle, the nitrogen cycle, the phosphorus cycle, or others—have a common structure, sharing three basic components: inputs, internal cycling, and outputs.

Input of Nutrients
The input of nutrients to an ecosystem depends on the type of biogeochemical cycle. Nutrients with a gaseous cycle, such as carbon and nitrogen, enter an ecosystem from the atmosphere. For example, carbon enters ecosystems almost solely through photosynthesis, which converts carbon dioxide to organic carbon compounds. Nitrogen enters ecosystems through a few pathways including lightning, nitrogen-fixing bacteria, and atmospheric deposition. In agricultural ecosystems, nitrogen fertilization provides a great amount of nitrogen influx, much larger than by any other influx pathways.

In contrast to carbon and nitrogen with input from the atmosphere, the input of nutrients such as calcium and phosphorus depends on the weathering of rocks and minerals. Soil characteristics and the process of soil formation have a major influence on processes involved in nutrient release to recycling pools. Supplementary soil nutrients come from airborne particles and aerosols, as wet or dry depositions. Such atmospheric deposition can supply more than half of the input of nutrients to some ecosystems.

The major sources of nutrients for aquatic ecosystems are inputs from the surrounding land. These inputs can take the forms of drainage water, detritus and sediment, and precipitation. Flowing aquatic systems are highly dependent on a steady input of detrital material from the watershed through which they flow.

Internal Cycling
Internal cycling of nutrients occurs when nutrients are transformed in ecosystems. Plants take up mineral (mostly inorganic) nutrients from soil through their roots and incorporate them into living tissues. Nutrients

in the living tissues occur in various forms of organic compounds and perform different functions in terms of physiology and morphology. When these living tissues reach senescence, the nutrients are usually returned to the soil in the form of dead organic matter. However, nitrogen can be reabsorbed from senescent leaves and transferred to other living tissues. Various microbial decomposers transform the organic nutrients into mineral forms through a process called mineralization. The mineralized nutrients are once again available to the plants for uptake and incorporation into new tissues. This process is repeated, forming the internal cycle of nutrients. Within the internal cycles, the majority of nutrients are stored in organic forms, either in living tissues or dead organic matter, whereas mineral nutrients represent a small proportion of the total nutrient pools.

Output of Nutrients

The output of nutrients from an ecosystem represents a loss. Output can occur in various ways, depending on the nature of a specific biogeochemical cycle. Carbon is released from ecosystems to the atmosphere in the form of carbon dioxide via the process of respiration by plants, animals, and microorganisms. Nitrogen is lost to the atmosphere in gaseous forms of nitrogen, nitrous oxide, and ammonia, mostly as by-products of microbial activities in soil. Nitrogen is also lost through leaching from the soil and carried out of ecosystems by groundwater flow to streams. Leaching also results in export of carbon, phosphorus, and other nutrients out of ecosystems.

Output of nutrients from ecosystems can also occur through surface flow of water and soil erosion.

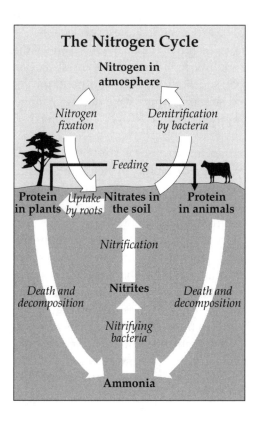

The Nitrogen Cycle

Nitrogen in atmosphere

Nitrogen fixation

Denitrification by bacteria

Feeding

Protein in plants *Uptake by roots* **Nitrates in the soil** **Protein in animals**

Nitrification

Death and decomposition **Nitrites** *Death and decomposition*

Nitrifying bacteria

Ammonia

However, loss of nutrients from one ecosystem may represent input to other ecosystems. Output of organic matter from terrestrial ecosystems constitutes the majority of nutrient input into stream ecosystems. Organic matter can also be transferred between ecosystems by herbivores. For example, moose feeding on aquatic plants can transport nutrients to adjacent terrestrial ecosystems and deposit them in the form of feces.

Considerable quantities of nutrients are lost permanently from ecosystems by harvesting, especially in farming and logging lands, when biomass is directly removed from ecosystems. Fire usually results in the loss of large amounts of nutrients. Fire kills vegetation and converts portions of biomass and organic soil matter to ash. Fire causes loss of nutrients through volatilization and airborne particulate. After fire, many nutrients become readily available, and nutrients in ash are subject to rapid mineralization. If not taken up by plants during vegetation recovery, nutrients are likely to be lost from ecosystems through leaching and erosion.

The Hubbard Brook Example

Nutrient cycling has been studied in several intact ecosystems. One of the most notable experiments was conducted in the Hubbard Brook experimental forest in New Hampshire. The experimental forest was established initially for forest hydrology research. Begun in the early 1960's, one of the longest-running studies of water and nutrient dynamics of forest ecosystems has been on the Hubbard Brook site. Both water and nutrient concentrations in precipitation inputs and stream outputs were regularly monitored, allowing estimations of nutrient balances over the watershed ecosystems.

One of the major findings from the Hubbard Brook study was that undisturbed forests exhibit regularity and predictability in their input-output balances for water and certain chemical elements. Nitrogen, however, shows a more complex, but still explicable, pattern of stream concentrations. Losses of nitrates from the control watershed are higher in the dormant season, when biological activity is low. Losses are near zero during the growing season, when biological demand for nitrogen by plants and microbes are high. Removal of vegetation in the Hubbard Brook forest had a marked effect on water and nutrient balances. Summer stream flow during the devegetation experiment was nearly four times higher than in the control watershed. The increase in stream flow, combined with increases in the concentration of nutrients within the stream, resulted in increases in loss rates of nitrate much higher than those of undisturbed areas. Similarly, loss of potassium used in large quantities by plants showed the greatest increase.

Nutrient Uptake and Competition

Ecosystem nutrient cycling is critical for plant growth and ecosystem productivity. Plant uptake of essential nutrient elements is related to nutrient availability, root absorption surface, rooting depth, and uptake kinetics of roots. A nutrient-rich site usually supports more plants of different species than a site with fewer available nutrients. Nutrient competition among plants is usually manifested through physiological, morphological, and ecological traits. Usually grasses and forbs can coexist in one grassland ecosystem, for example, through different rooting depth. To compete for less soluble nutrients such as phosphorus, plants usually extend their root surfaces using symbiotic relationships with mycorrhizae. Differential seasonality in nutrient uptake and rooting depth become more critical to compete for limited nutrients.

Yiqi Luo

See also: Balance of nature; Biomass related to energy; Competition; Food chains and webs; Geochemical cycles; Herbivores; Hydrologic cycle; Omnivores; Phytoplankton; Rain forests and the atmosphere; Trophic levels and ecological niches.

Sources for Further Study

Aber, J. D., and J. M. Melillo. *Terrestrial Ecosystems.* 2d ed. San Diego: Academic Press, 2001.

Likens, G. E., et al. *Biogeochemistry of a Forested Ecosystem.* New York: Springer-Verlag, 1977.

Schlesinger, W. H. *Biogeochemistry: An Analysis of Global Change.* 2d ed. San Diego: Academic Press, 1997.

OCEAN POLLUTION AND OIL SPILLS

Types of ecology: Aquatic and marine ecology; Ecotoxicology

Oil spills resulting from human error often affect marine and coastal areas. Past oil spills in different areas of the world demonstrate that environmental damage depends on the toxicity and the persistence of the oil; both vary widely depending on a variety of factors.

Why Oil Spills Occur

Almost all imported and Alaskan oil is transported to U.S. refineries and consumers by oceangoing tankers. Most oil spills result from marine transportation accidents, with human error usually playing a major role. Navigation errors, equipment malfunctions, bad judgment, and even the inability of all crew members to speak a common language have all been major contributing factors in the largest and most environmentally damaging oil spills. Not all oil spills are environmental disasters, but spilled oil can decimate plant and animal populations by a combination of mechanical toxicity and chemical toxic effects resulting from an organism's physiological reaction to the chemicals present in oil.

Environmental Damage

The most common sight during an oil spill is dark, gelatinous masses of "mousse"—an oil and water emulsion that floats on the water, sticking to everything with which it comes into contact. Mousse usually causes the majority of the environmental damage during an oil spill by the process of mechanical toxicity, as it suffocates and smothers organisms that ingest it or are covered by it. Seabirds and furry marine mammals are highly susceptible to this process, succumbing to exposure, dehydration, or starvation.

Crude oil is a complex mixture of thousands of different chemicals called hydrocarbons, named after a molecular structure based on hydrogen and carbon atoms. Different hydrocarbons vary in their chemical properties, toxicity, and behavior during an oil spill. The major groups are classified by molecular geometry and weight. The low-molecular-weight molecules (aliphatics) are single-bonded, chain-shaped molecules, such as gasoline. They are the most chemically reactive and volatile, and they are acutely toxic. These compounds tend to evaporate or burn easily during an oil spill and therefore do not persist in the environment for long periods. Intermediate-molecular-weight hydrocarbons, or aromatics, are ring-shaped molecules, such as benzene. They are also highly reactive and cause biological impacts because of both acute and chronic toxicity. Aromatic hydrocarbon compounds are more environmentally persistent than aliphatics. Since many are carcinogens, they can cause different forms of biological damage, disease, and death even after a long time period and in low doses. The high-molecular-weight oil compounds are mostly polycyclic aromatic hydrocarbons structured of ring shapes bonded together to form molecules. Although they are not very chemically reactive and do not dissolve well in water, many are carcinogenic. They tend to be very environmentally persistent.

For hundreds of millions of years before humans evolved, oil was "spilled" naturally into the world's oceans by natural oil seeps—fractures in the earth's crust that tap deep, oil-bearing rocks. A variety of natural processes act to reduce the environmental impacts of this oil, and these same processes also take place during a human-caused oil spill. Oil is dispersed from the oil slick and into the larger environment by five basic processes. Evaporation of the low-molecular-weight hydrocarbon compounds removes most of the oil relatively quickly. Sunlight can degrade additional oil in a process called photodegradation if the oil is exposed for enough time. Because oil is an organic substance, additional oil is removed by natural biodegradation thanks to "oil-eating" microorganisms. Most of the rest of the oil either washes up onto a coastal area or breaks up into heavy "tar balls" rich in high-molecular-weight hydrocarbons that eventually sink.

Some oil spills put so much oil into the environment that these processes cannot respond quickly enough to prevent environmental damage. Other factors can also enhance environmental damage from oil spills. Some types of oil or refined petroleum products are more toxic than others. Oil spills in cold climates generally cause more damage because cold temperatures retard evaporation and the microbial metabolic rates necessary for rapid oil removal. Furthermore, sunlight is often of low intensity, which retards photodegradation. Wave conditions and tidal currents can affect how much oil washes up onto a coastal area and how rapidly it is moved elsewhere or removed. Finally, the amount of environmental damage from an oil spill is highly dependent on the type of coastal environment oiled in the spill, as coastal environments vary in density (or biomass) and varieties of wildlife. Coastlines also vary in the degree to which they are sheltered from natural oil-removal processes. In general, rocky headlands, wave-cut rock platforms, and reefs exposed to high wave activity suffer far less damage during an oil spill than do sheltered marshes, tidal flats, and mangrove forests. The damage on beaches is related to the grain size of the beach sediment. Fine-sand beaches are relatively flat and hard-packed, and oil does not soak into the sediment or persist for long. Oil will soak deeply into coarse sand, gravel, and shell beaches, causing more damage over a longer period.

Most of what has been learned about oil spill behavior, environmental damage, and oil spill cleanup techniques comes from studying past spills. In most cases, spill prevention is far cheaper and more effective than spill response, and cleanup efforts usually capture very little of the spilled oil.

Ixtoc I

The Ixtoc I spill of June 3, 1979, was the result of an explosion, or "blowout," of an offshore oil well that was drilling into a subsurface oil reservoir. Although human error was definitely a factor, the cause of the blowout remains unresolved. It has been blamed on the use of drilling mud that was not dense enough to counteract the pressure of the oil and gas at depth, as well as on the improper installation of the blowout preventer—a fail-safe device used on drilling rigs to prevent just this type of disaster. The result was a continuous 290-day oil spill, during which an estimated 475,000 metric tons of crude oil (one metric ton equals approximately five barrels) were released into the environment. In addition to doing considerable environmental damage on the coast of Mexico, oil fouled much of the barrier island coast of Texas. However, most of the oil did not make it to shore, and the final accounting for this spill gives a good indication of the long-term fate of spilled oil in offshore areas: 1 percent burned at the spill site, 50 percent evaporated, 13 percent photodegraded or biodegraded, 7 percent washed up on the coast (6 percent in Mexico, 1 percent in Texas), 5 percent was mechanically removed by skimmers and booms, and 24 percent sank to the seafloor (assumed by mass balance).

The *Exxon Valdez*

The *Exxon Valdez* oil spill—which occurred in Prince William Sound, Alaska, on March 24, 1989—is a good example of how environmental damage follows human error and inadequate response. After departing Port Valdez with a full cargo, the *Exxon Valdez* oil tanker struck a well-charted submerged rock reef located 1.6 kilometers outside the shipping lane. The ship was under the command of an unlicensed third mate in calm seas and left the shipping lane with permission from the Coast Guard to avoid ice. However, it strayed too close to the reef before evasive action was attempted. The captain was under the influence of alcohol during events leading to the

Seabirds are often casualties of ocean oil spills, succumbing to exposure, dehydration, starvation, suffocation, or the oil's toxicity. In 1989, experts estimated that, in addition to thousands of sea otters and deer, more than half a million seabirds died as a result of the Exxon Valdez *oil spill in Prince William Sound, Alaska.* (PhotoDisc)

accident. His blood alcohol level nine hours after the grounding was measured at 0.06 percent; the estimate at the time of the accident was 0.19 percent—almost twice the legal level for drivers in California. Convicted of negligence and stripped of his commander's license, he was subsequently employed as an instructor to teach others to operate supertankers.

Leaking oil was observed immediately. Oil-spill response crews funded by Exxon and the Alyeska Pipeline Consortium—oil companies that used the Port Valdez terminal—were poorly prepared and reacted too slowly and with inadequate equipment. The first response arrived ten hours after the accident with insufficient booms and skimmers. Chemical dispersants applied to break up the oil slick were ineffective in the calm seas and caused the oil slick to thin and spread more rapidly. Four days later, the weather changed: 114-kilometer-per-hour winds mixed the oil with seawater, creating a frothy mousse. More than 65,000 metric tons of oil spilled over several weeks. About 15,600 square kilometers of ocean and 1,300 kilometers of shoreline were affected.

Federal estimates of wildlife mortality include 3,500 to 5,500 otters; 580,000 seabirds; and 300 deer poisoned by eating oiled kelp. Economic damages totaled more than $5 billion. The long-term effects on commercial marine organisms, larval organisms, and bottom-dwelling life are not known.

Exxon promised to clean nearly 500 kilometers of shoreline by September, 1989, but cleaned only 2 kilometers during the first month after the spill. Exxon and its contractors used a variety of cleanup techniques, including placing booms and skimmers, sopping up oil with absorbent materials, scraping oil by hand from rocks, stimulating the growth of oil-eating bacteria cultures, and washing coastal areas with cold water, hot water, and steam. The use of hot water and steam was effective at cosmetically removing surface oil, but it did not remove oil that had soaked into the sediment; the technique killed most organisms that had escaped the oil. The oil washed from the beach was to be collected by booms and skimmers offshore, but this process was so inefficient that much of the oil migrated to tide pools that had not been affected by the spill directly. Ironically, only eighteen months after the spill, life had returned to oiled coasts that had received little or no cleanup, while beaches cleaned with hot water were still relatively sterile and required several years to repopulate. Exxon announced that it would not return to clean more shoreline in 1990 but relented under threat of a court order from the Coast Guard to enforce federal cleanup requirements. During the summer of 1990, shoreline cleanup resumed, including application of fertilizer to stimulate growth of naturally occurring oil-eating bacteria, a technique that is not very efficient in the cold waters of southern Alaska.

The tale of the *Exxon Valdez* is not complete without mentioning that the Port Valdez Coast Guard did not have state-of-the-art radar equipment for monitoring ship movement in this heavily used and environmentally sensitive area. In the early 1980's, federal and state funds for monitoring the Port Valdez oil companies' compliance with oil-spill preparedness legislation had been cut by more than 50 percent. The original environmental impact statement for oil-handling activity in Prince William Sound included an agreement that defines cleanup responsibility for oil spills. Exxon, as the company responsible for the spill, was to pay the first $14 million of cleanup costs, with $86 million in additional cleanup funds from the Alyeska contingency fund. Thus, the maximum financial responsibility to oil companies from a spill was $100 million unless the spill was judged to be caused by negligence. Cleanup activities ceased eighteen months after the spill with total expenditure of $2.2 billion; most of this at taxpayer expense. In 1994, a federal court unanimously awarded $5.3 billion in punitive and compensatory damages, the largest-ever jury award, to some thirty-five thousand people impacted by the spill. By June, 1999, Exxon had yet to pay a single dollar as the case continued through the legal process. Finally, it is interesting to note that Exxon's estimate of cleanup costs in late 1989 were $500 million, and it carried $400 million of oil spill liability insurance. Exxon saved $22 million by not building the *Exxon Valdez* with a double hull; its 1988 annual profits were $5,300 million.

According to National Oceanic and Atmospheric Administration (NOAA) estimates, less than 1 percent of *Exxon Valdez*'s oil burned at the site, 20 percent evaporated, 8 percent was mechanically removed, and nearly 72 percent was deposited on the seafloor. According to Exxon's estimates, 7 percent of the oil burned at the site, 32 percent evaporated, 9 percent photodegraded or biodegraded, 15 percent was mechanically removed, and 37 percent was assumed deposited on the seafloor.

Operation Desert Storm

The February, 1991, Operation Desert Storm oil spill—the largest oil spill in history—occurred when the Iraqi military opened valves and pumps at Sea Island Terminal, a tanker loading dock located 16 kilometers off the coast of Kuwait. This facility has a production capacity of 100,000 barrels per day, about three *Exxon Valdez* loads each week. The Iraqis also opened plugs on five Kuwaiti tankers, spilling an additional 60,000 barrels. The estimate for the entire spill is 6 million barrels, or roughly 30 times the volume of the *Exxon Valdez*. About 650 square kilometers of coast were heavily contaminated.

Three days after starting the spill, the Iraqis ignited the oil leaking from the terminal. This was the best thing to happen from an environmental perspective. During most spills, more oil is removed by natural evaporation than by any cleanup technique; igniting the oil merely speeds up this process. Burning can be an important mechanism for removing oil from the sea and avoiding environmental damage, and tests have shown purposeful ignition in open water away from the coast to be an excellent oil-slick fighting strategy. However, this must be done within the first few hours of the spill—in order to maintain the fire, the slick must be more than 1 millimeter thick and must contain relatively little emulsified water. To maintain thickness, the slick is best surrounded with fireproof booms. However, at the time of the Operation Desert Storm spill, almost all the fireproof booms in the world were in Prince William Sound. Saudi Arabia also used dispersants on portions of the slick, but this effort was too late to be effective before a thick mousse had formed.

The prime objectives of causing the spill were to hamper an amphibious military landing by oiling the beaches and to disrupt desalinization of drinking water at Khafji and Jubail, the two primary sources of potable water for Saudi Arabia. The Saudis used booms to protect the plant intakes with great success. The retreating Iraqis also ignited more than seven hundred of about one thousand inland wells, resulting in an additional 6 million barrels per day burned. This volume eventually made the marine spill insignificant, and the burning created 3 percent of total global carbon emissions during the time period of the event.

The Arabian Gulf is an unusual body of water. It is very shallow (average 33 meters) and is nearly enclosed as a marine basin. Because it is also microtidal (the tidal range is less than 0.6 meter), it flushes out slowly (once every two hundred years, compared with once every few days for Prince William Sound). It is also important to remember that this is not a pristine marine environment. Natural oil seeps are very common, there is a general lack of environmental standards and poor cooperation among Persian Gulf nations, and virtually no oil spill preparations or equipment were present in this part of the world. Earlier spills had occurred in the region, but they typically were associated with ongoing wars; the hostile environment made it difficult to utilize spill abatement specialists and equipment. For example, during the Iran-Iraq War, Iraq attacked an Iranian offshore platform (Nowruz) in 1982, spilling more than 2 million barrels of oil, a volume nearly half as large as the Operation Desert Storm spill. Losses of marine mammals and birds were great, and populations had not rebounded by the time of Operation Desert Storm.

During the Operation Desert Storm oil spill, about 180 kilometers of Saudi Arabian coastline were oiled (65 kilometers were severely damaged), and oil reached south as far as the United Arab Emirates and Bahrain. Much of the southern Kuwaiti coast (sea-grass beds, marsh, and mangroves) was severely damaged, and about 25 percent of the Saudi shrimp industry was lost. Although some twenty thousand wading birds were killed, no deaths of dolphins or dugongs were reported. However, these animals suffered greatly during the Iran-Iraq War. Estimates of the time required for ecological renewal of the Persian Gulf following the Operation Desert Storm spill (one to four years) were relatively short for two reasons: The high water temperature results in high microbial activity and biodegradation of the oil, and much of the oil was burned.

James L. Sadd

See also: Acid deposition; Lakes and limnology; Marine biomes; Pollution effects; Reefs.

Sources for Further Study
Alaska Wilderness League. *Preventing the Next Valdez: Ten Years After Exxon's Spill New Disasters Threaten Alaska's Environment.* Washington, D.C.: Alaska Wilderness League, 1999.

Etkin, Dagmar Schmidt. *Financial Costs of Oil Spills in the United States.* Arlington, Mass.: Cutter Information, 1998.

_____. *Marine Spills Worldwide.* Arlington, Mass.: Cutter Information, 1999.

Hall, M. J. *Crisis on the Coast.* Portland, Oreg.: USCG Marine Safety Office, 1999.

Smith, Roland. *The Sea Otter Rescue: The Aftermath of an Oil Spill.* New York: Puffin Books, 1999.

OLD-GROWTH FORESTS

Types of ecology: Biomes; Ecosystem ecology; Restoration and conservation ecology

Ancient ecosystems, old-growth forests consist of trees that have never been harvested. These forests are, in some cases, the only habitat for a number of plant and animal species.

The timber industry views large, old trees as a renewable source of fine lumber, but environmentalists see them as part of an ancient and unique ecosystem that can never be replaced. In the 1970's scientists began studying the uncut forests of the Pacific Northwest and the plants and animals that inhabited them. In a U.S. Forest Service publication, *Ecological Characteristics of Old-Growth Douglas-Fir Forests* (1981), Forest Service biologist Jerry Franklin and his colleagues showed that these forests were not just tangles of dead and dying trees but rather a unique, thriving ecosystem made up of living and dead trees, mammals, insects, and fungi.

Old-Growth Forest Ecosystem

The forest usually referred to as old growth occurs primarily on the western slope of the Cascade Mountains in southeast Alaska, southern British Columbia, Washington, Oregon, and Northern California. The weather there is wet and mild, ideal for the growth of trees such as Douglas fir, cedar, spruce, and hemlock. Studies have shown that there is more biomass, including living matter and dead trees, per acre in these forests than anywhere else on earth. Trees can be as tall as 300 feet (90 meters) with diameters of 10 feet (3 meters) or more and can live as long as one thousand years. The forest community grows and changes over time, not reaching biological climax until the forest primarily consists of hemlock trees, which are able to sprout in the shade of the sun-loving Douglas fir.

One of the most important components of the old-growth forest is the large number of standing dead trees, or snags, and fallen trees, or logs, on the forest floor and in the streams. The fallen trees rot very slowly, often taking more than two hundred years to completely decompose. During this time they are important for water storage, as wildlife habitat, and as "nurse logs" where new growth can begin. In fact, seedlings of some trees, such as western hemlock and Sitka spruce, have difficulty competing with the mosses on the forest floor and need to sprout on the fallen logs.

Another strand in the complex web of the forest consists of mycorrhizal fungi (mycorrhizae), which attach themselves to the roots of the trees and enhance their uptake of water and nutrients. The fruiting bodies of these fungi are eaten by small mammals such as voles, mice, and chipmunks, which then spread the spores of the fungi in their droppings. There are numerous species of plants and animal wildlife that appear to be dependent on this ecosystem to survive.

Protecting the Forest
By the 1970's most of the trees on timber industry-owned lands had been cut. Their replanted forests, known as second growth, would not be ready for harvest for several decades, so the industry became increasingly dependent on public lands for their raw materials. Logging of old growth in the national forests of western Oregon and Washington increased from 900 million board feet in 1946 to more than 5 billion board feet in 1986.

Environmentalists claimed that only 10 percent of the region's original forest remained. Determined to save what was left, they encouraged the use of the evocative term "ancient forest" to counteract the somewhat negative connotations of "old growth." Then they were given an effective tool in the northern spotted owl. This small bird was found to be dependent on

Old-growth forests in the Pacific Northwest are the only habitat for not only the famous endangered northern spotted owl but also a host of other wildlife species. (PhotoDisc)

453

old growth, and its listing under the federal Endangered Species Act in 1990 caused a decade of scientific, political, and legal conflict.

Under law, the U.S. Forest Service was required to protect enough of the owl's habitat to ensure its survival. An early government report identified 7.7 million acres of forest to be protected for the bird. Later, the U.S. Fish and Wildlife Service recommended 11 million acres. In 1991 U.S. District Court judge William Dwyer placed an injunction on all logging in spotted owl habitat until a comprehensive plan could be finalized. The timber industry responded with a prediction of tens of thousands of lost jobs and regional economic disaster. In 1993 President Bill Clinton convened the Forest Summit conference in Portland, Oregon, to work out a solution. The Clinton administration's plan, though approved by Judge Dwyer, satisfied neither the industry nor the environmentalists, and protests, lawsuits, and legislative battles continued.

As the twentieth century came to an end, timber harvest levels had been significantly reduced, the Northwest's economy had survived, and additional values for old-growth forests were found: habitat for endangered salmon and other fish, a source for medicinal plants, and a repository for benefits yet to be discovered. The decades-long controversy over the forests of the Northwest had a deep impact on environmental science as well as natural resource policy and encouraged new interest in other native forests around the world, from Brazil to Malaysia to Russia.

Joseph W. Hinton

See also: Endangered animal species; Forests; Habitats and biomes; Lakes and limnology; Mountain ecosystems; Rain forests; Rain forests and the atmosphere; Savannas and deciduous tropical forests; Slash-and-burn agriculture; Taiga; Tundra and high-altitude biomes.

Sources for Further Study

Dietrich, William. *The Final Forest: The Battle for the Last Great Trees of the Pacific Northwest.* New York: Penguin Books, 1993.

Durbin, Kathie. *Tree Huggers: Victory, Defeat, and Renewal in the Northwest Ancient Forest Campaign.* Seattle: Mountaineers, 1996.

Kelly, David, and Gary Braasch. *Secrets of the Old Growth Forest.* Salt Lake City, Utah: Gibbs-Smith, 1988.

Maser, Chris. *Forest Primeval: The Natural History of an Ancient Forest.* Corvallis: Oregon State University Press, 2001.

OMNIVORES

Types of ecology: Behavioral ecology; Ecoenergetics

Omnivores are animals that eat both plants and animals. They are found in all types of animals, including arthropods, fish, birds, and mammals. Omnivore diets may vary seasonally.

Many animals are either herbivores, which eat only plant food, or carnivores, which eat only the flesh of other animals. The preference for one type of food or the other depends largely on the type of digestive system that the animal has, and the resources it can put into its "energy budget."

Ecological Advantages

Meat is generally easier to digest and requires a less complex digestive system and a relatively short intestinal tract. However, in order to get meat, carnivores have to invest a lot of time hunting their prey, and the outcome of a hunt is always uncertain. The food of herbivores is much easier to obtain, since plants do not move and all the herbivore has to do is graze on the grasses, leaves, or algae readily available around it. However, the cellulose that plants are made of is very tough to digest, and thus herbivores must have a much lengthier and more complex digestive tract than carnivores. Many herbivores are ruminants, with multipart stomachs, which have to chew and digest their food more than once in order to get adequate nutrition from it.

Carnivores and herbivores are also vulnerable to a loss of their food source. Herbivores whose digestive systems are specialized to process only one type of food will starve if that food becomes scarce as a result of drought or some other climatic change. Carnivores often have specialized hunting patterns that cannot be changed if the prey (usually herbivores) become scarce due to loss of their own food source.

Omnivores maximize their ability to obtain food by having digestive tracts capable of processing both plant and animal food, although they are usually not capable of digesting the very tough plant material, such as grasses and leaves, that many large herbivores eat. Omnivores may also be scavengers, eating whatever carrion they may come across. Omnivores often lack the specialized food-gathering ability characteristic of pure carnivores and herbivores. Many animals often thought of as carnivores are actually omnivores, eating both plants and animals.

Diversity of Omnivores

Omnivores can be found among all types of animals, living on land and in water. They include fish, mollusks, arthropods, birds, and mammals.

Most insects are either herbivores, such as grasshoppers, or carnivores such as mantises. However some, such as yellow jacket wasps, are omnivores, eating other insects, fruit, and nectar. Omnivorous snails and slugs eat algae, leaves, lichens, insects, and decaying plant and animal matter. Their main organ for eating is called a radula, a tonguelike, toothed organ that is drawn along rocks, leaves, or plants to scrape off food; it is also used to bore holes through shells of other mollusks, to get to their flesh.

Omnivorous fish include the common carp, goldfish, catfish, eels, and minnows. Since a fish's food is often suspended in the medium through which the fish swims—water—being able to gulp up whatever comes into its mouth is an efficient way for a fish to eat. Similarly, bottom-feeders (fish that suck up material from the floor of whatever body of water they inhabit) also benefit from not needing to sort through the material before they ingest it.

Many birds are omnivores, such as robins, ostriches, and flamingos. The pink or red color of flamingos occurs because they eat blue-green algae and higher plants which contain the same substances that make tomatoes red. They also eat shrimp and small mollusks.

Mammal omnivores include bears; members of the weasel family, such as skunks; the raccoon family (raccoons and coatimundis); monkeys; apes; and humans. Raccoons and coatis, found only in the Americas, eat insects, crayfish, crabs, fish, amphibians, birds, small mammals, nuts, fruits, roots, and plants. Like other omnivores, they also eat carrion. Bears eat grass, roots, fruits, insects, fish, small or large mammals, and carrion.

Sanford S. Singer

See also: Balance of nature; Biomass related to energy; Food chains and webs; Herbivores; Nutrient cycles; Predation; Trophic levels and ecological niches.

Sources for Further Study

Kay, Ian. *Introduction to Animal Physiology*. New York: Springer-Verlag, 1999.

Lauber, Patricia. *Who Eats What?* New York: HarperTrophy, 1995.

Llamas, Andreu. *Crustaceans: Armored Omnivores*. Milwaukee: Gareth Stevens, 1996.

McGinty, Alice B. *Omnivores in the Food Chain*. Logan, Iowa: Powerkids Press, 2002.

OZONE DEPLETION AND OZONE HOLES

Types of ecology: Ecotoxicology; Global ecology

Ozone occurs naturally in the atmosphere and absorbs ultraviolet radiation from the sun. In the past few decades, a "hole" in the atmosphere's ozone layer has been recorded over Antarctica, and its size, although fluctuating, has increased over time. Scientists are concerned with the damage to all living organisms from ultraviolet radiation if the atmosphere's ozone continues to decrease.

Ozone in the Atmosphere

Ozone, although only a minor component of the atmosphere, plays a vital role in the survival of life on earth. Ozone molecules absorb high-energy ultraviolet (UV) radiation, which humans perceive as light, from the sun. Absorption of ultraviolet radiation in the ozone layer, a region of the stratosphere that contains the maximum concentration of ozone, prevents most such light from reaching the surface of the planet. If none of the sun's ultraviolet radiation were blocked by the ozone layer, it would be difficult for most forms of life, including humans, to survive on land.

The concentration of ozone in the atmosphere is highly variable, changing with altitude, geographic location, time of day, time of year, and prevailing local atmospheric conditions. Long-term fluctuations in ozone concentration are also seen, some of which are related to the solar sunspot cycle. While long-term average ozone concentrations are relatively stable, short-term fluctuations of as much as 10 percent in total column abundance of ozone as a result of the natural variability in ozone concentration are often observed.

Discovery of a "Hole"

Beginning in the early 1970's, a new and unexpected decrease in stratospheric ozone concentration was first observed. The decrease was localized in geography to the Southern Polar region, and in time to early spring (which begins in October in the Southern Hemisphere). The initial decrease in ozone was small, but by 1980, decreases in total column abundance of ozone of as much as 30 percent were being seen, well outside the range of variation expected as a result of random fluctuations. This seasonal destruction of stratospheric ozone above Antarctica, which by 1990 had reached 50 percent of the total column abundance of ozone, was soon given the label "ozone hole."

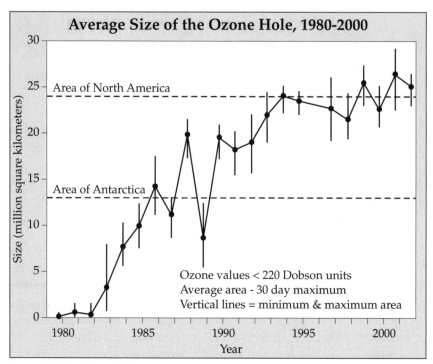

Source: Goddard Space Flight Center. National Aeronautics and Space Administration. http:\\toms.gsfc.nasa.gov\multi\oz_hole_area.jpg; accessed April 15, 2002.

The Role of CFCs

While it was initially unclear whether formation of the Antarctic ozone hole stemmed from natural causes or from anthropogenic effects on the environment, extensive field studies combined with the results of laboratory experiments and computer modeling of the atmosphere quickly led to a consistent and detailed explanation for ozone-hole formation. The formation of the ozone hole has two principal causes: chemical reactions that occur generally throughout the stratosphere, and special conditions that exist in the Antarctic region.

Under normal conditions, the concentration of ozone in the stratosphere is determined by a balance between reactions that remove ozone and those that produce ozone. The removal reactions are mainly catalytic chain reactions, in which trace atmospheric chemical species destroy ozone molecules without themselves being consumed. In such processes, it is possible for one chain carrier to remove many ozone molecules before being itself removed. The trace species involved in ozone removal include hydrogen oxides and nitrogen oxides, formed primarily by naturally occurring processes, and chlorine and bromine atoms and their corresponding oxides.

A major source of chlorine in the stratosphere is the decomposition of a class of compounds called chlorofluorocarbons (CFCs). Such compounds can be used in refrigeration and air conditioning, as aerosol propellants, and as solvents. Chlorofluorocarbons are extremely stable in the lower atmosphere, with lifetimes of several decades. The main fate of chlorofluorocarbons is slow migration into the stratosphere, where they absorb ultraviolet light and release chlorine atoms. The chlorine atoms produced from the breakdown of chlorofluorocarbons in the stratosphere provide an additional catalytic process by which stratospheric ozone can be destroyed. A similar set of reactions involving a class of bromine-containing compounds called halons, used in some types of fire extinguishers, leads to additional ozone destruction. By 1986, the average global loss of stratospheric ozone caused by the release of chlorofluorocarbons, halons, and related compounds into the environment was estimated to be 2 percent.

The Antarctic Stratosphere

While the decomposition and subsequent reaction of chlorofluorocarbons and other synthetic compounds explains the small general decline in ozone concentration observed in the stratosphere, additional processes are needed to account for the more massive seasonal ozone depletion observed above Antarctica. These processes involve a set of special conditions that in combination are unique to the stratosphere above Antarctica.

During daylight hours, a portion of the chlorine present in the stratosphere is tied up in the form of reservoir species, compounds such as hydrogen chloride and chlorine nitrate that do not react with ozone. This slows the rate of removal of ozone by chlorine. Processes that directly or indirectly involve absorption of sunlight transform reservoir species into ozone-destroying chlorine atoms. During the Antarctic winter, when sunlight is entirely absent, stratospheric chlorine is rapidly converted into reservoir species.

In the absence of additional chemical processes, the onset of spring in Antarctica and the return of sunlight would convert a portion of the reservoir compounds into reactive chlorine species and reestablish the balance between ozone-producing and ozone-destroying processes. However, the extremely low temperatures occurring in the stratosphere above Antarctica during the winter months leads to the formation of polar stratospheric clouds, which, because of the extremely low concentration of water vapor in the stratosphere, do not form during other seasons or outside the polar regions of the globe. The ice crystals that compose the clouds act as catalysts that convert reservoir species into diatomic chlorine and other gaseous chlorine compounds that, in the presence of sunlight, re-form ozone-destroying species. At the same time, the nitrogen oxides found in the reservoir species are converted into nitric acid, which remains attached to the ice crystals. As these ice crystals are slowly removed from the stratosphere by gravity, the potential for conversion of active forms of chlorine into reservoir species is greatly reduced. Because of this, when spring arrives, large amounts of ozone-destroying chlorine species are produced by the action of sunlight, and only a small fraction of this reactive chlorine is converted into reservoir species. The increased rate of ozone removal caused by the abundance of reactive chlorine present in the stratosphere leads to ozone depletion and formation of the ozone hole.

An additional process important in formation of the ozone hole is the unique air-circulation pattern in the stratosphere above Antarctica. During the winter and early spring, a vortex of winds circulates about the South Pole. This polar vortex minimizes movement of ozone and reservoir-forming compounds from other regions of the stratosphere. As this polar vortex breaks up in midspring, ozone concentrations in the Antarctic stratosphere return to normal levels, and the ozone hole gradually disappears.

Study and Interpretation
Researchers utilize a great diversity of devices and techniques in their study and interpretation of atmospheric ozone. One popular technique is the use of models. A good model is one that simulates the interrelationships and interactions of the various parts of the known system. The weakness of models is that, often, not enough is known to give an accurate picture of the total system or to make accurate predictions. Most modeling is done on computers. Scientists estimate how fast chemicals such as CFCs and nitrous oxide will be produced in the future and build a computer model of the way these chemicals react with ozone and with one another. From this model, it is possible to estimate future ozone levels at different altitudes and at different future dates.

Arctic Depletion?
Similar processes appear to be at work in the Arctic stratosphere, leading to ozone depletion, as in the Antarctic; however, the National Oceanic and Atmospheric Administration (NOAA) Aeronomy Laboratory in Boulder, Colorado, reported a discrepancy between observed ozone depletion and predicted levels, based on models that account accurately for the Antarctic depletions. This report suggests that some other mechanism is at work in the Arctic. Thus, good models can be very useful in studying new data.

There are two models favored by most scientists in this area. Some scientists put forth a chemical model that says the depletion is caused by chemical events promoted by the presence of chlorofluorocarbons created by industrial processes. Acceptance of this model was promoted by the discovery of fluorine in the stratosphere. Fluorine does not naturally occur there, but it is related to CFCs. The other model assumes that the ozone hole was formed by dynamic air movement and mixing. This model best fits data gathered by ozone-sensing balloons that sample altitudes up to 30 kilometers and then radio the data back to Earth. Ozone depletion is confined to air between 12 and 20 kilometers. While the total ozone depletion is 35 percent, different strata showed various amounts of depletion from 70 to 90 percent. Surprisingly, about half the ozone was gone in twenty-five days. This finding does not fit the chemical model very well.

Besides ozone-sensing balloons, satellites are of much help. The National Aeronautics and Space Administration (NASA) obtains measurements with its Nimbus 7 satellite. Ozone measurements made by this satellite helped to develop flight plans for the specialized aircraft NASA also deploys in ozone studies. NASA's ER-2 aircraft is a modified U-2 reconnaissance plane that carries instruments up to 20 kilometers in altitude for seven-hour flights to 80 degrees north latitude. A DC-8, operating during the same period, is able to survey the polar vortex, owing to its greater range. In addition, scientists utilize many meteorological techniques and instruments, including chemical analysis of gases by means of infrared spectroscopy, mass spectroscopy and gas spectroscopy combined, gas chromatography, and oceanographic analysis of planktonic life in the southern Atlantic, Pacific, and Indian oceans. As new research methods become available, they are applied to this essential study.

Public Health Concerns

Atmospheric ozone provides a gauze of protection from the lethal effects of ultraviolet radiation from the sun. This ability to absorb ultraviolet radiation protects all life-forms on the earth's surface from excessive ultraviolet radiation, which destroys the life of plant and animal cells. Currently, between 10 and 30 percent of the sun's ultraviolet B (UV-B) radiation reaches the earth's surface. If ozone levels were to drop by 10 percent, the amount of UV-B radiation reaching Earth would increase by 20 percent.

Present-day UV-B levels are responsible for the fading of paints and the yellowing of window glazing and for car finishes becoming chalky. These kinds of degradation will accelerate as the ozone layer is depleted. There could also be increased smog, urban air pollution, and a worsening of the problem of acid rain in cities. In humans, UV-B causes sunburn, snow blindness, skin cancer, cataracts, and excessive aging and wrinkling of skin. Skin cancer is the most common form of cancer—more than 400,000 new cases are reported every year in the United States alone. The National Academy of Sciences has estimated that each 1 percent decline in ozone would increase the incidence of skin cancer by 2 percent. Therefore, a 3 percent depletion in ozone would produce some 20,000 more cases of skin cancer in the United States every year.

Ecological Concerns
Many other forms of life—from bacteria to forests and crops—are adversely affected by excessive radiation as well. Ultraviolet radiation affects plant growth by slowing photosynthesis and by delaying germination in many plants, including trees and crops. Scientists have a great concern for the organisms that live in the ocean and the effect ozone depletion may have on them. Phytoplankton, zooplankton, and krill (a shrimplike crustacean) could be greatly depleted if there were a drastic increase in ultraviolet A and B. The result would be a tremendous drop in the population of these free-floating organisms. These organisms are important because they are the beginning of the food chain. Phytoplankton use the energy of sunlight to convert inorganic compounds, such as phosphates, nitrates, and silicates, into organic plant matter. This process provides food for the next step in the food chain, the herbivorous zooplankton and krill. They, in turn, become the food for the next higher level of animals in the food chain. Initial studies of this food chain in the Antarctic suggest that elevated levels of ultraviolet radiation impair photosynthetic activity. Recent studies show that a fifteen-day exposure to UV-B levels 20 percent higher than normal can kill off all anchovy larvae down to a depth of 10 meters. There is also concern that ozone depletion may alter the food chain and even cause changes in the organism's genetic makeup. An increase in the ultraviolet radiation is likely to lower fish catches and upset marine ecology, which has already suffered damage from human-made pollution. On a worldwide basis, fish presently provides 18 percent of all the animal protein consumed.

International Response
The United Nations Environmental Program (UNEP) is working with governments, international organizations, and industry to develop a framework within which the international community can make decisions to minimize atmospheric changes and the effects they could have on the earth. In 1977, UNEP convened a meeting of experts to draft the World Plan of Action on the Ozone Layer. The plan called for a program of re-

search on the ozone layer and on what would happen if the layer were damaged. In addition, UNEP created a group of experts and government representatives who framed the Convention for the Protection of the Ozone Layer. This convention was adopted in Vienna in March, 1985, by twenty-one states and the European Economic Community and has subsequently been signed by many more states. The convention pledges states that sign to protect human health and the environment from the effects of ozone depletion. Action has already been taken to protect the ozone layer. Several countries have restricted the use of CFCs or the amounts produced. The United States banned the use of CFCs in aerosols in 1978. Some countries, such as Belgium and the Nordic countries, in effect banned CFC production altogether.

George K. Attwood and Jeffrey A. Joens

See also: Acid deposition; Deforestation; Global warming; Greenhouse effect; Pollution effects; Rain forests and the atmosphere.

Sources for Further Study

Bast, Joseph L., Peter J. Hill, and Richard C. Rue. *Eco-Sanity: A Common Sense Guide to Environmentalism.* Lanham, Md.: Heartland Institute, 1994.

Cagin, Seth, and Philip Dray. *Between Earth and Sky: How CFCs Changed Our World and Endangered the Ozone Layer.* New York: Pantheon Books, 1993.

Firor, John. *The Changing Atmosphere: A Global Challenge.* New Haven, Conn.: Yale University Press, 1990.

Fisher, David E. *Fire and Ice: The Greenhouse Effect, Ozone Depletion, and Nuclear Winter.* New York: Harper & Row, 1990.

Graedel, T. E., and Paul J. Crutzen. *Atmospheric Change: An Earth System Perspective.* New York: W. H. Freeman, 1993.

Roan, Susan. *Ozone Crisis: The Fifteen-Year Evolution of a Sudden Global Emergency.* New York: John Wiley & Sons, 1989.

Rowland, F. Sherwood. "Stratospheric Ozone Depletion." *Annual Review of Physical Chemistry* 42 (1991): 731.

Shell, E. R. "Weather Versus Chemicals." *The Atlantic* 259 (May, 1987): 27-31.

Somerville, Richard C. J. *The Forgiving Air: Understanding Environmental Change.* Los Angeles: University of California Press, 1996.

PALEOECOLOGY

Types of ecology: Evolutionary ecology; Paleoecology

Paleoecology is the study of ancient organisms and their relationships to one another and to their environments. The characteristics of ancient environments may be determined by examining rock and fossil features.

As a field of science, paleoecology is most closely related to paleontology, the study of fossils. It is also related to paleoclimatology, paleogeography, and a number of other areas of study dealing with the distant past. All these disciplines have a handicap in common: Because they deal with the past, scientists are unable to apply the usual scientific criteria of direct observation and measurement of phenomena. Therefore, they look to a number of different methods whereby evidence of past conditions and organisms can be deduced.

Dendrochronology

One of the most intensively investigated paleoecological problems has been the changing environments associated with the ice ages of the past million years. Analysis of pollen from bogs in many parts of the world indicates that there have been at least four advances and retreats of glaciers during that period. Evidence for this is the changing proportions of pollen from tree species found at the various depths of bogs. In North America, for example, spruces (indicators of cool climate) formerly lived much farther south than they do now. They were largely replaced almost eight thousand years ago by other tree species, such as oaks, which are indicative of warmer climates. This warming trend was a result of the latest glacial retreat.

Tree-ring analysis, also known as dendrochronology, not only enables paleoecologists to date past events such as forest fires and droughts but also allows them to study longer-term cycles of weather and climate, especially those of precipitation and temperature. In addition, trees serve as accumulators of past mineral levels in the atmosphere and soil. Lead levels of tree wood showed a sharp increase as the automobile became common in the first half of the twentieth century because of lead additives in gasoline. Tree rings formed since the 1970's have shown a decrease in lead because of the decline in use of leaded fuels. Tree-ring analysis has also been a valuable tool for archaeologists' study of climatic changes responsible for shifting patterns of population and agriculture among native Americans of the southwestern United States.

The Fossil Record

Fossil evidence is the chief source of paleoecological information. A fossil bed of intact clam shells with both valves (halves) present in most individuals, for example, usually indicates that the clams were preserved in the site in which they lived (called autochthonous deposition). Had they been transported by currents or tides to another site of deposition (allochthonous deposition), the valves would have been separated, broken, and worn. Similarly, many coal beds have yielded plant fossils that indicate that their ancient environments were low-lying swamp forests with sluggish drainage periodically flooded by water carrying a heavy load of sand. The resulting fossils may include buried tree stumps and trunks with roots still embedded in their original substrate and numerous fragments of twigs, leaves, and bark within the sediment.

Certain dome- or mushroom-shaped structures called stromatolites are found in some of the most ancient of earth's sedimentary rocks. These structures may be several meters in diameter and consist of layers of material trapped by blue-green algae (cyanobacteria). Such structures are currently being formed in shallow, warm waters. Uniformitarian interpretation of the three-billion-year-old stromatolites is that they were formed under similar conditions. Their frequent association with mud cracks and other shallow- and above-water features leads to the interpretation that they were formed in shallow inshore environments subject to frequent exposure to the air.

Relative oceanic temperature can be estimated by observing the direction in which the shells of certain planktonic organisms coil. The shell of *Globigerina pachyderma* coils to the left in cool water and to the right in warmer water. *Globigerina menardii* shells coil in an opposite fashion—to the right in cool water and to the left in warmer water. Uniformitarian theory leads one to believe that ancient *Globigerina* populations responded to water temperature in a similar manner. Sea-bottom core samples showing fossils with left- or right-coiling shells may be used to determine the relative water temperature at certain periods. Eighteen-thousand-year-old sediments taken from the Atlantic Ocean show a high frequency of left-handed *pachyderma* and right-handed *menardii* shells. Such observations indicate that colder water was much farther south about eighteen thousand years ago, a date that corresponds to the maximum development of the last Ice Age.

Fossil Deposition

Fossil arrangement and position can be a clue to the environments in which the organisms lived or in which they were preserved. Sea-floor currents can align objects such as small fish and shells. Not only can the existence of the current be inferred, but also its direction and velocity can be determined. Currents and tides can create other features in sediments which are sometimes indicators of environment. If a mixture of gravel, sand, silt, and clay is being transported by a moving body of water such as a stream, tide, or current, the sediments will often become sorted by the current and be deposited as conglomerates—sandstones, siltstones, and shales. Such graded bedding can be used to determine the direction and velocity of currents. Larger particles, such as gravel, would tend to be deposited nearer the sediment source than smaller particles such as clay. Similarly, preserved ripple marks indicate current direction. Mud cracks in a rock layer indicate that the original muddy sediment was exposed to the atmosphere at least for a time after its deposition.

Fossil Composition

Certain minerals within fossil beds or within the fossil remains themselves can sometimes be used to interpret the paleoenvironment. The presence of pyrite in a sediment almost always indicates that the sedimentary environment was deficient in oxygen, and this, in turn, often indicates deep, still water. Such conditions exist today in the Black Sea and even in some deep lakes, with great accumulations of dead organic matter.

The method of preservation of the remains of the fossilized organism can be an indication of the environment in which the creature lived (or died).

Amber, a fossilized resin, frequently contains the embedded bodies of ancient insects trapped in the resin like flies on flypaper. This ancient environment probably contained resin-bearing plants (mostly conifers), and broken limbs and stumps that oozed resin to trap these insects. Mummified remains in desert areas and frozen carcasses in the northern tundra indicate the environments in which the remains were preserved thousands of years ago.

Marks made on fossil parts by other organisms offer indirect evidence of the presence and activity of other species that might not have left fossil remains. Predators and scavengers can leave such marks on bones and shells by boring, scratching, and gnawing. One of the most controversial taphonomic problems in paleoecology is distinguishing between tooth marks left by animal scavengers and predators on bones and those marks left by the stone and bone tools of early human ancestors.

Fossil Assemblages and Trace Fossils

Fossil assemblages (thanatocoenoses) are the most commonly used indicators of ancient environments. The use of any fossil in interpreting the past must be subject to several qualifications. The fossil record is sparse for most groups of organisms because fossilization itself is a relatively rare event. Rapid burial of the remains and the presence of hard body parts (wood, shells, bones, and teeth) are only two of several fossilization prerequisites that must usually be met. This means that terrestrial organisms and soft-bodied organisms are seldom fossilized. Events leading to fossilization after the death of an organism (taphonomy) usually destroy the soft tissues through decay and scavenging and often disrupt and distort the remaining hard parts through transportation and weathering.

An additional taphonomic problem is encountered when clumps or clusters of fossil remains are located. Without careful study, it is difficult to determine whether these assemblages are truly representative of the groupings of the organisms in life or if they are simply coincidental aggregations of such items as shells and limbs that were swept together by currents or wind and thus not indicative of the living situation and environment. Because of limitations on the interpretation of ancient environments by the use of fossilized body parts, trace fossils are often more reliable indicators of environmental conditions. Trace fossils are preserved tracks, burrows, trails, and other indirect indications of the presence of an organism. The presence of marine worm burrows, for example, can indicate environmental factors such as salinity and depth. Such traces are not transported from one site to another. Transportation results in their destruction. Whenever these imprints are found, therefore, paleoecologists are able to make some inferences about the environment in which they were formed.

Stratigraphy

One of the most important methods to be mastered by paleoecologists is stratigraphy, the science of correlating and determining the age of rock layers with those of the fossils contained within these layers or formations. Rock layers or strata are not usually connected over large regions. While they might have been deposited as sediments at the same time and under the same conditions, subsequent erosion has usually made the layers discontinuous. Stratigraphers attempt to correlate discontinuous rock strata by measuring and describing them and by noting the presence of unique fossils called index fossils. If two strata are correlated, then they were probably deposited during approximately the same period, although there may be a gradation of conditions.

For example, there may be a layer of sediment deposited at the same time, but under nearshore conditions at one spot and under offshore conditions at another. Relative ages are determined by using the law of superposition: Older rocks lie beneath younger rocks. One can say that a certain stratum is older than, the same age as, or younger than another layer, depending upon their relative positions. Absolute ages (estimated age in years before the present) are determined by measuring the amounts of certain radioactive elements within igneous rocks. Such radiometric age determinations are of less value for sedimentary rocks since they give the age of the minerals of the rock, not the age of the rock itself.

Related Fields

Paleoecological data are applicable to other, related paleo-fields of the earth and life sciences. The study of fossils, paleontology, is enhanced by the inclusion of information about the fossil organisms' environments and relationships with other organisms. Paleontologists should attempt to reconstruct ancient environments because organisms did not exist alone or in vacuums: They lived in dynamic biological communities. Paleogeography relies heavily on paleoecological information to discern the locations, directions, and time intervals of glaciation, deposition of sediments, temperature, and other environmental variables. This information has been used to determine the past positions of continents and has been a valuable contribution to scientists' knowledge of continental drift.

Paleoclimatologists, who study ancient regional and planetwide conditions, must make use of local bits of paleoecological information to see the big picture of climate. One of the major concerns of paleoclimatology is the recognition of planetary climatic cycles and associated environmental and biological cycles. If there is a repeated recurrence of global environmental change, then predictions about future climatic change become more accurate and probable.

P. E. Bostick

See also: Adaptations and their mechanisms; Adaptive radiation; Coevolution; Colonization of the land; Convergence and divergence; Dendrochronology; Evolution: definition and theories; Evolution: history; Evolution of plants and climates; Extinctions and evolutionary explosions; Natural selection; Nonrandom mating, genetic drift, and mutation; Punctuated equilibrium vs. gradualism; Speciation; Species loss.

Sources for Further Study

Agashe, Shripad N. *Paleobotany: Plants of the Past, Their Evolution, Paleoenvironment, and Application in Exploration of Fossil Fuels*. Enfield, N.H.: Science Publishers, 1997.

Arduini, Paolo, and Giorgio Teruzzi. *Simon and Schuster's Guide to Fossils*. New York: Simon & Schuster, 1986.

Bennett, K. D. *Evolution and Ecology: The Pace of Life.* New York: Cambridge University Press, 1997.

Brett, Carlton E., and Gordon C. Baird, eds. *Paleontological Events: Stratigraphic, Ecological, and Evolutionary Implications.* New York: Columbia University Press, 1997.

Cowen, Richard. *History of Life.* 3d ed. Malden, Md.: Blackwell Science, 2000.

Davis, Richard A. *Depositional Systems: A Genetic Approach to Sedimentary Geology.* Englewood Cliffs, N.J.: Prentice-Hall, 1983.

Dodd, J. Robert, and Robert J. Stanton, Jr. *Paleoecology: Concepts and Applications.* 2d ed. New York: John Wiley & Sons, 1990.

National Research Council. Commission on Geosciences, Environment, and Resources. Board on Earth Sciences and Resources. *Effects of Past Global Change on Life.* Washington, D.C.: National Academy Press, 1995.

Newton, Cathryn, and Léo Laporte. *Ancient Environments.* 3d ed. Englewood Cliffs, N.J.: Prentice-Hall, 1989.

Shipman, Pat. *Life History of a Fossil: An Introduction to Taphonomy and Paleoecology.* Cambridge, Mass.: Harvard University Press, 1993.

PESTICIDES

Types of ecology: Agricultural ecology; Ecotoxicology

Pesticides are substances designed to kill unwanted plants, fungi, or animals that interfere, directly or indirectly, with human activities. The unintended impacts of pesticides such as DDT have been to change ecosystems and their components.

The major types of pesticides in common use are insecticides (to kill insects), nematocides (to kill nematodes), fungicides (to kill fungi), herbicides (to kill weeds), and rodenticides (to kill rodents). Herbicides and insecticides make up the majority of the pesticides applied in the environment. Biopesticides are beneficial microbes, fungi, insects, or animals that kill pests. While the use of pesticides has mushroomed since the introduction of monoculture (the agricultural practice of growing only one crop on a large amount of acreage), the application of toxins to control pests is by no means new.

Insecticides: History of Use

The use of sulfur as an insecticide dates back before 500 B.C.E. Salts from heavy metals such as arsenic, lead, and mercury were used as insecticides from the fifteenth century until the early part of the twentieth century, and residues of these toxic compounds are still being accumulated in plants that are grown in soil where these materials were used. In the seventeenth and eighteenth centuries, natural plant extracts, such as nicotine sulfate from tobacco leaves and rotenone from tropical legumes, were used as insecticides. Other natural products, such as pyrethrum from the chrysanthemum flower, garlic oil, lemon oil, and red pepper, have long been used to control insects.

In 1939 the discovery of dichloro-diphenyl-trichlorethane (DDT) as a strong insecticide opened the door for the synthesis of a wide array of synthetic organic compounds to be used as pesticides. Chlorinated hydrocarbons such as DDT were the first group of synthetic pesticides. Other commonly used chlorinated hydrocarbons have in the past included aldrin, endrin, lindane, chlordane, and mirex. Because of their low biodegradability and persistence in the environment, they proliferated up the food chain and became concentrated in predators, such as birds and animals butchered for human consumption. The use of these compounds was therefore banned or severely restricted in the United States, but only after years of use.

Organophosphates such as malathion, parathion, and methamidophos have replaced the chlorinated hydrocarbons. These compounds biodegrade in a fairly short time but are generally much more toxic to humans and other animals than the compounds they replaced. In addition, they are water-soluble and therefore more likely to contaminate water supplies. Carbamates such as carbaryl, maneb, and aldicarb have also been used in place of chlorinated hydrocarbons. These compounds rapidly biodegrade and are less toxic to humans than organophosphates, but they are less effective in killing insects.

Herbicides

Herbicides are classified according to their method of killing rather than their chemical composition. As their name suggests, contact herbicides such as atrazine and paraquat kill when they come in contact with a plant's leaf surface. Contact herbicides generally disrupt the photosynthetic mechanism. Systemic herbicides such as diuron and fenuron circulate throughout the plant after being absorbed. They generally mimic the plant hormones and cause abnormal growth to the extent that the plant can no longer supply sufficient nutrients to support growth. Soil sterilants such as triflurain, diphenamid, and daiapon kill microorganisms necessary for plant growth and also act as systemic herbicides.

The spraying of herbicides and pesticides remains a standard agricultural practice in the cultivation of both food crops and ornamental crops such as these tulips. (PhotoDisc)

471

Current Use

In the United States, approximately 55,000 different pesticide formulations are available, and Americans apply about 500 million kilograms (1.1 billion pounds) of pesticides each year. Fungicides account for 12 percent of all pesticides used by farmers, insecticides account for 19 percent, and herbicides account for 69 percent. These pesticides have been used primarily on four crops: soybeans, wheat, cotton, and corn. Approximately $5 billion is spent each year on pesticides in the United States, and about 20 percent of this is for nonfarm use. On a per-unit-of-land basis, homeowners apply approximately five times as much pesticide as do farmers. On a worldwide basis, approximately 2.5 tons (2,270 kilograms) of pesticides are applied each year. Most of these chemicals are applied in developed countries, but the amount of pesticide used in developing countries is rapidly increasing. Approximately $20 billion is spent worldwide each year, and this expenditure is expected to increase in the future, particularly in the developing countries.

Despite current concerns about their toxicity and biomagnification, pesticide use has had a beneficial impact on the lives of humans by increasing food production and reducing food costs. Even with pesticides, pests reduce the world's potential food supply by as much as 55 percent. Without pesticides, this loss would be much higher, resulting in increased starvation and higher food costs. Pesticides also increase the profit margin for farmers. It has been estimated that for every dollar spent on pesticides, farmers experience an increase in yield worth three to five dollars. Pesticides appear to work better and faster than alternative methods of controlling pests. These chemicals can rapidly control most pests, are cost-effective, can be easily shipped and applied, and have a long shelf life compared to alternative methods. In addition, farmers can quickly switch to another pesticide if genetic resistance to a given pesticide develops.

Perhaps the most compelling argument for the use of pesticides is the fact that pesticides have saved lives. It has been suggested that since the introduction of DDT, the use of pesticides has prevented approximately seven million premature human deaths from insect-transmitted diseases such as sleeping sickness, bubonic plague, typhus, and malaria. Perhaps even more lives have been saved from starvation because of the increased food production resulting from the use of pesticides. It has been argued, therefore, that this one benefit outweighs the potential health risks of pesticides. In addition, new pesticides are continually being developed, and safer and more effective pest control may be available in the future.

The publication of Rachel Carson's classic Silent Spring *(1962), which outlined the ecotoxicity of the pesticide DDT, helped stimulate the environmental movement of the 1960's.* (Library of Congress)

Environmental Concerns

In spite of all the advantages of using pesticides, their benefit must be balanced against the potential environmental damage they can cause. An ideal pesticide would have the following characteristics: It should not kill any organism other than the target pest; it would in no way affect the health of nontarget organisms; it would degrade into nontoxic chemicals in a relatively short time; it would prevent the development of resistance in the organism it is designed to kill; and it would be cost-effective. Since no currently available pesticide meets all of these criteria, a number of environmental problems have developed, one of which is broad-spectrum poisoning. Most, if not all, chemical pesticides are not selective; they kill a wide range of organisms rather than just the target pest. Killing beneficial insects, such as bees, lady bird beetles, and wasps, may result in a range of problems. For example, reduced pollination and explosions in the populations of unaffected insects can occur.

When DDT was first used as an insecticide, many people believed that it was the final solution for controlling many insect pests. Initially, DDT dramatically reduced the number of problem insects; within a few years, however, a number of species had developed genetic resistance to the chemical and could no longer be controlled with it. By the 1990's there were approximately two hundred insect species with genetic resistance to DDT. Other chemicals were designed to replace DDT, but many insects also developed resistance to these newer insecticides. As a result, although many synthetic chemicals have been introduced to the environment, the pest problem is still as great as it ever was.

Depending on the type of chemical used, pesticides remain in the environment for varying lengths of time. Chlorinated hydrocarbons, for example, can persist in the environment for up to fifteen years. From an economic standpoint, this can be beneficial because the pesticide has to be applied less frequently, but from an environmental standpoint, it is detrimental. In addition, when many pesticides are degraded, their breakdown products, which may also persist in the environment for long periods of time, can be toxic to other organisms.

Pesticides may concentrate as they move up the food chain, a process called biomagnification. All organisms are integral components of at least one food pyramid. While a given pesticide may not be toxic to species at the base, it may have detrimental effects on organisms that feed at the apex because the concentration increases at each higher level of the pyramid. With DDT, for example, some birds can be sprayed with the chemical without any apparent effect, but if these same birds eat fish that have eaten insects that contain DDT, they lose the ability to metabolize calcium properly. As a result, they lay soft-shelled eggs, which causes deaths of most of the offspring.

Pesticides can be hazardous to human health. Many pesticides, particularly insecticides, are toxic to humans, and thousands of people have been killed by direct exposure to high concentrations of these chemicals. Many of these deaths have been children who were accidentally exposed to toxic pesticides because of careless packaging or storage. Numerous agricultural laborers, particularly in developing countries where there are no stringent guidelines for handling pesticides, have also been killed as a result of direct exposure to these chemicals. Workers in pesticide factories are also a high-risk group, and many of them have been poisoned through job-related contact with the chemicals. Pesticides have been suspected of causing long-term health problems such as cancer. Some of the pesticides have been shown to cause cancer in laboratory animals, but there is currently no direct evidence to show a cause-and-effect relationship between pesticides and cancer in humans.

D. R. Gossett

See also: Biomagnification; Biopesticides; Genetically modified foods; Integrated pest management; Soil contamination.

Sources for Further Study
Altieri, Miguel A. *Agroecology: The Science of Sustainable Agriculture.* 2d ed. Boulder, Colo.: Westview Press, 1995.

Carson, Rachel. *Silent Spring*. 1962. Reprint. Thorndike, Maine: G. K. Hall, 1997.

Mannion, Antionette M., and Sophia R. Bowlby. *Environmental Issues in the 1990's*. Chichester, N.Y.: John Wiley & Sons, 1992.

Milne, George W. A., ed. *Ashgate Handbook of Pesticides and Agricultural Chemicals*. Burlington, Vt.: Ashgate, 2000.

Nadakavukaren, Anne. *Man and Environment: A Health Perspective*. 2d ed. Prospect Heights, Ill.: Waveland Press, 1990.

Pierce, Christine, and Donald VanDeVeer. *People, Penguins, and Plastic Trees*. 2d ed. Belmont, Calif.: Wadsworth, 1995.

PHEROMONES

Types of ecology: Behavioral ecology; Chemical ecology; Physiological ecology

Pheromones are chemicals or mixtures of chemicals that are used as messages between members of a species. They are integral parts of the social communication within most species. They may prove to be of great value in pest control and in enhancing agricultural production.

Pheromones are chemical signals. Originally defined to include only signals between individuals of the same animal species, the term has been generalized to designate any chemical or chemical mixture that, when released by one member of any species, affects the physiology or behavior of another member of the same species. "Pheromone" is also one of a set of terms developed to express the chemical interactions in ecological communities.

All pheromones are semiochemicals that carry information between members of a single species. To do so, the pheromone must be released into the atmosphere or placed on some structure in the organism's environment. It is thus made available to other members of the species for interpretation and response. It is also available to members of other species, however, so it is a potential allelochemical.

Types of Pheromones

There are two general types of pheromone: those that elicit an immediate and predictable behavioral response, called releaser or signal pheromones, and those that bring about a less obvious physiological response, called primer pheromones (because they prime the system for a possible behavioral response). Pheromones are also categorized according to the messages they carry. There are trail, marker, aggregation, attractant, repellant, arrestant, deterrent, stimulant, alarm, and other pheromones. Their functions are suggested by the terms used to name them.

To appreciate fully the complexity of the interactions under consideration, it is important to remember that a pheromone may also be acting as a kairomone, allomone, or hormone. For example, klipspringer antelope mark vegetation in their environment with a chemical secreted from a special gland. Other klipspringer investigate the marks to gather information on the marking individual. Ticks that parasitize the klipspringer, however, are also attracted to the chemical marks and thus increase their chance of attaching to a host when the mark is renewed or when another klipspringer investigates it. The tick is using the pheromone as a kairomone. Pheromones can act as allomones as well, though the interaction is sometimes less direct. Bolas spiders produce the sex-attractant pheromone of a female moth and use it to lure male moths to a trap. The spider uses the moth pheromone as an allomone.

Pheromonal Compounds and Strategies

The chemical compounds that act as pheromones are numerous and diverse. Most are lipids or chemical relatives of the lipids, including many steroids. Even a single pheromonal message may require a number of different compounds, each present in the proper proportion, so that the active pheromone is actually a mixture of chemical compounds.

Different physical and chemical characteristics are required for pheromones with different functions. Attractant pheromones must generally be volatile to permit atmospheric dispersal to their targets. Many female insects emit sex-attractant pheromones to advertise their readiness to mate. The more widely these can be dispersed, the more males the advertisement will reach. On the other hand, many marking pheromones need not be especially volatile because they are placed at stations which are checked periodically by the target individuals. The klipspringer marking pheromone is an example. Some pheromones are exchanged by direct contact, and these need not have any appreciable volatile component. Many mammals rub, lick, and otherwise contact one another in social contexts and exchange pheromones at these times.

Specificity also varies for pheromones with different functions. Sex attractants usually need to be very specific, directed only to members of the opposite sex and the same species. Alarm pheromones, on the other hand, need not be so specific. These pheromones simply alert other members of the same species to a disturbance. It is usually harmless, and sometimes even helpful, to alert members of other species as well. In keeping with this argument, related groups of ant species produce species-specific sex-attractant pheromones: Each female attracts only males of its own species. In contrast, alarm pheromones of any species in the group will stimulate defensive reactions in individuals of many in the species.

Pheromonal systems are not organized in any standard way in different species. Many mites and ticks also have nonspecific alarm pheromones. Surprisingly, some groups also have nonspecific sex-attractant pheromones. In these cases, the specificity necessary for reproductive efficiency is generated by species-specific mating stimulant pheromones. These pheromones are produced by a female after males have been attracted to her. They stimulate mating behavior, but only in males of the same species. Thus the required specificity is achieved by a different mechanism. This is only one of many examples of the diversity of pheromonal schemes among organisms.

Pheromone Sources and Receptors

The sources of pheromones are also diverse. Some pheromones are produced by specialized glands; many insect species have glands specialized for the production of pheromones. One example is the harvester ant's alarm pheromone, which is produced in the mandibular gland at the base of the jaws. Other pheromones seem to be by-products of other bodily functions. The lipids of mammalian skin are probably primarily important in waterproofing and in maintaining the outer layer of the skin, but many also function as pheromones.

The reproductive tract is an important source of pheromones in many species. These usually act as sex-attractant or sex-stimulant pheromones or as signal pheromones that give information on the sexual state of the emitter. The urine and feces of many species also contain pheromones that are used to mark territory boundaries and to transmit other information about the marking individual. Many pheromones seem to be produced not by the sending organism alone but by microorganisms living on the skin or in the glands or cavities of the sender's body. These microbes convert products of their host into the actual signal molecules, or pheromones, used by the host.

The receptors for pheromones are also of many different types, and the chemical receptors for taste and smell are often involved. In vertebrates, the vomeronasal organ (Jacobson's organ) seems to be an important receptor for many pheromones. It is a pouch off the mouth or nasal passages, and it contains receptors similar to those for smell. It is nonfunctional in humans, but it functions in more primitive mammals and seems to be of great importance to snakes and other reptiles. Insects and other invertebrates have many specialized structures for receiving pheromonal messages. Perhaps the best-known example is the feathery antennae of many male moths, which are receptors for the female moth's sex-attractant pheromone. Some pheromones seem to be absorbed through the skin or internal body linings and to bring about their effects by attaching to some unknown internal receptor.

Prevalence in Nature

Pheromones are widespread in nature, occurring in most, if not all, species. Most are poorly understood. The best-known are those found in insects, partly because of their potential use in the control of pest populations and partly because the relative simplicity of insect behavior allowed for rapid progress in the identification of pheromones and their actions. Despite these advantages, much remains to be learned even about insect pheromones. Mammalian pheromones are not as well known, although they may also be of economic importance. The more complex behavior of mammals makes the study of their responses to pheromones much more difficult.

Research Methods

Both behavioral and chemical techniques are required to study pheromones and other semiochemicals. The observation of behavior, either in nature or in captivity, often suggests pheromonal functions. These hypothesized functions are then tested by presenting the pheromone to a potentially responsive organism and observing the response. Situations may be arranged which demand the subject's response to a particular pheromone under otherwise natural conditions. Alternatively, the organisms may be observed in enclosures to help control the experimental context. The presentation of the hypothetical pheromone may be in the form of another organism of the same species or some structure to which the presumed pheromone has been applied. The observed response (or lack of response) gives information on the status of the presented chemical as a pheromone in that behavioral context.

While the pheromonal function of secretions from a gland or other source can be determined from these behavioral tests, the tests can give information on specific chemical compounds only if the compounds can be isolated and identified. The isolation and identification of pheromonal compounds are challenging because of the great complexity of the secretions in which they are found and the exceptionally small amounts that are required to elicit a response. Many separation and identification techniques are used. One of the most powerful is a combination of gas chromatography and mass spectrometry.

Gas chromatography is used to separate and sometimes to identify chemicals that are volatile or can be made volatile. The unknown chemical is mixed with an inert gas, called the mobile phase of the gas chromatography system. This mixture is passed through a tube containing a solid, called the stationary phase. The inert gas does not interact with the solid; however, many of the compounds mixed with it do, each to an extent determined by the characteristics of the compound and the characteristics of the stationary phase. Some members of the mixture will interact very strongly with the solid and so move slowly through the tube, whereas others may not interact with the solid at all and so pass through rapidly. Other members of the mixture interact at intermediate strengths and so spend intermediate amounts of time in the tube. The different compounds are recorded and collected separately as they exit from the tube.

For identification, the compounds are often passed on to a mass spectrometer. In mass spectrometry the compound is broken up into electrically charged particles. The particles are then separated according to their mass-to-charge ratio, and the relative number of particles of each mass-to-charge ratio is recorded and plotted. The original compound can usually be identified by the pattern produced under the specific conditions used. After separation and identification, the individual chemicals may be subjected to behavioral studies.

Uses for Pheromones
Pheromones and other semiochemicals are of interest simply from the standpoint of understanding communication between living things. In addition, they have the potential to provide effective, safe agents for pest control. The possibilities include sex-attractant pheromones to draw pest insects of a particular species to a trap (or to confuse the males and keep them from finding females) and repellant pheromones to drive a species of insect away from a valuable crop species. One reason for the enthusiasm generated by pheromones in this role is their specificity. Whereas insecticides generally kill valuable insects as well as pests, pheromones will often be specific for one or a few species.

These chemicals were presented as a panacea for insect and other pest problems in the 1970's, but most actual attempts to control pest populations failed. Many people in the field have suggested that lack of understanding of the particular pest and its ecological context was the most common cause of failure. They maintain that pest-control applications must be made with extensive knowledge and careful consideration of pest characteristics and the ecological system. In this context, pheromones have become a part of integrated pest management (IPM) strategies, in which they are used along with the pest's parasites and predators, resistant crop varieties, insecticides, and other weapons to control pests. In this role, pheromones have shown great promise.

Some consideration has been given to the control of mammalian pests with pheromones, though this field is not as well developed as that of insect control. Pheromonal control of mammalian reproduction has received considerable attention for other reasons: Domestic mammals are of great economic importance, and many wild mammalian species are endangered to the point that captive breeding has been attempted. The manipulation of reproductive pheromones may be used to enhance reproductive potentials in both cases. The complexity of mammalian behavioral and reproductive systems, however, and the subtle changes brought about by mammalian pheromones present a particular challenge. As with insect pest control, the key to progress is a complete understanding of the entire system being manipulated.

Pheromones and other semiochemicals are of great potential economic importance as substitutes for or adjuncts to toxic pesticides in pest management. Mammalian reproductive pheromones are being explored as tools to enhance reproductive efficiency in domestic and endangered mammals. A complete understanding of the complex roles of pheromones in each of the systems being managed is necessary for success in all these endeavors.

Carl W. Hoagstrom

See also: Allelopathy; Biopesticides; Communication; Defense mechanisms; Displays; Ethology; Insect societies; Isolating mechanisms; Metabolites; Reproductive strategies.

Sources for Further Study

Agosta, William C. *Chemical Communication: The Language of Pheromones.* New York: Scientific American Library, 1992.

Albone, Eric S. *Mammalian Semiochemistry: The Investigation of Chemical Signals Between Mammals.* New York: John Wiley & Sons, 1984.

Booth, William. "Revenge of the 'Nozzleheads.' " *Science* 239 (January 8, 1988): 135-137.

Carde, Ring T., and Albert K. Minks, eds. *Insect Pheromone Research: New Directions.* New York: Chapman and Hall, 1997.

Mayer, Marion S., and John R. McLaughlin. *Handbook of Insect Pheromones and Sex Attractants.* Boca Raton, Fla.: CRC Press, 1991.

Mitchell, Everett R., ed. *Management of Insect Pests with Semiochemicals: Concepts and Practice.* New York: Plenum, 1981.

Mittler, Thomas E., Frank J. Radovsky, and Vincent H. Resh, eds. *Annual Review of Entomology.* Vol. 46. Palo Alto, Calif.: Annual Reviews, 2000.

Nordlund, Donald A., Richard L. Jones, and W. Joe Lewis, eds. *Semiochemicals: Their Role in Pest Control.* New York: John Wiley & Sons, 1981.

Vandenbergh, John G., ed. *Pheromones and Reproduction in Mammals.* New York: Academic Press, 1983.

Wilson, Edward O. "Pheromones." *Scientific American* 206 (May, 1963): 100-114.

PHYTOPLANKTON

Types of ecology: Ecoenergetics; Ecotoxicology

Most plankton are microscopic and are usually single-celled, a chain of cells, or a loose group of cells. Algal and cyanobacterial plankton are referred to as phytoplankton.

The term "plankton," from Greek *planktos* for "wandering," is applied to any organism that floats or drifts with the movement of the ocean water. Whereas the heterotrophic crustaceans and larvae of animals are referred to as zooplankton, phytoplankton phytoplankton (literally, "plant" plankton) refer to a collection of diverse, largely algal and cyanobacterial, microorganisms. The phytoplankton include diatoms, unicellular cyanobacteria and coccolithophorids in nutrient-poor waters, and cryptomonads. They manufacture organic material from carbon dioxide, usually through photosynthesis, and therefore occupy the key trophic level of producers, at the base of the food chain. Phytoplankton are responsible for one-half of the world's primary photosynthesis and produce one-half of the oxygen in the atmosphere.

Eighty to ninety percent of the weight of phytoplankton is water, with the rest made up of protein, fat, salt, carbohydrates, and minerals. Some species have compounds of calcium or silica that make up their shells or skeletons. Phytoplankton include many of the algal phyla: *Chrysophyta* (chrysophytes), *Phaeophyta* (golden-brown algae), coccolithophores, silicoflagellates, and diatoms. The most common type of phytoplankton is the diatom (phylum *Bacillariophyta*), a single-celled organism that can form complex chains. Dinoflagellates (phylum *Dinophyta*) are the most complex of the phytoplankton. They are unicellular and mobile. Green algae (phylum *Chlorophyta*) are usually found in estuaries or lagoons in the late summer and fall. Some species can cause toxic algal blooms associated with coastal pollution and eutrophication. Cyanobacteria (often called bluegreen algae but not true algae) are prominent near shore waters with limited circulation and brackish waters.

Role in the Food Chain
Phytoplankton are primary producers, responsible for half the world's primary photosynthesis: the conversion of light energy and inorganic matter into bioenergy and organic matter. Each year, 28 billion tons of carbon and

250 billion to 300 billion tons of photosynthetically produced materials are generated in the oceans by phytoplankton. All animal organisms eliminate carbon dioxide into the atmosphere, and plants remove carbon dioxide from the air through photosynthesis. In the oceans' carbon cycle, carbon dioxide from the atmosphere dissolves in the ocean. Photosynthesis by marine plants, mainly phytoplankton, converts the carbon dioxide into organic matter. Carbon dioxide is later released by plants and animals during respiration, while carbon is also excreted as waste or in the dead bodies of organisms. Bacteria decompose organic matter and release the carbon dioxide back into the water. Carbon may be deposited as calcium carbonate in biogenous sediments and coral reefs (made of skeletons and shells of marine organisms).

Because they are primary producers of organic matter through photosynthesis, phytoplankton play a key role in the world's food chain: They are its very beginning. Sunlight usually penetrates only 200 to 300 feet deep into ocean waters, a region called the photic zone. Most marine plant and animal life and feeding take place in this zone. Phytoplankton, the first level in the marine food chain, are the primary food source for zooplankton and larger organisms. These microscopic plants use the sun's energy to absorb minerals to make basic nutrients and are eaten by herbivores, or plant eaters. Herbivores are a food source for carnivores, the meat eaters. In temperate zones, phytoplankton increase greatly in the spring, decline in the summer, and increase again in the fall. Zooplankton (animal plankton) are at their maximum abundance after the spring increase, and their grazing on the phytoplankton causes a decrease in phytoplankton population in the summer. Fish and invertebrates that eat zooplankton become more abundant and so on, up the food chain. Krill, planktonic crustaceans, and larvae commonly eaten by whales, fish, seals, penguins, and seabirds feed on diatom phytoplankton.

Red Tides

The term "red tide" is applied to red, orange, brown, or bright-green phytoplankton blooms, or even to blooms that do not discolor the water. Red tides are poorly understood and unpredictable. No one is certain what causes the rapid growth of a single species of phytoplankton, although they can blossom where sunlight, dissolved nutrient salts, and carbon dioxide are available to trigger photosynthesis. Dense phytoplankton blooms occur in stable water where lots of nutrients from sewage and run-off are available. Natural events, such as storms and hurricanes, may remobilize populations buried in the sediment. These nuisance blooms, usually caused by dinoflagellates, which turn the water a reddish brown, and cyanobacteria, are becoming more frequent in coastal waters, possibly because of increased human populations and sewage. In shallower bodies of water, such as bays and estuaries, nutrients from winter snow runoffs, spring rains, tributaries, and sewage bring about spring and summer blooms.

Some of the poisons produced during red tides are the most powerful toxins known. The release of toxins by dinoflagellates may poison the higher levels of the food chain as well as suppress other phytoplankton species. These toxins cause high mortality in fish and other marine vertebrates. They can kill the whales and seabirds that eat contaminated fish. Dinoflagellates produce a deadly neurotoxin called saxitoxin, which is fifty times more lethal than strychnine or curare. Commercial shellfish, such as mussels, clams, and crabs, can store certain levels of the toxin in their bodies.

People who eat contaminated shellfish may experience minor symptoms, such as nausea, diarrhea, and vomiting, or more severe symptoms such as loss of balance, coordination and memory, tingling, numbness, slurred speech, shooting pains, and paralysis. In severe cases, death results from cardiac arrest. When the toxins are blown ashore in sea spray, they can cause sore throats or eye and skin irritations.

Toxic blooms costs millions of dollars in economic losses, especially for fisheries which cannot harvest some species of shellfish. Smaller fish farms can be devastated. Additionally, coastal fish deaths foul beaches and shore water with decaying bodies, which can cripple tourism in the coastal regions.

Not all blooms are harmful, but they do affect the marine environment. Even when no toxins are released, massive fish kills can result when the large blooms of phytoplankton die. When the blooming phytoplankton population crashes, bacterial decomposition depletes the oxygen in the water, which in turn reduces water quality, and fish and other marine animals suffocate.

Virginia L. Hodges

See also: Biomass related to energy; Eutrophication; Food chains and webs; Marine biomes; Nutrient cycles; Trophic levels and ecological niches.

Sources for Further Study
Castro, Peter, and Michael Huber. *Marine Biology.* 3d ed. Boston: McGraw-Hill, 2000.
Cousteau, Jacques. *The Ocean World.* New York: Harry N. Abrams, 1985.

Levinton, Jeffrey S. *Marine Biology: Function, Biodiversity, Ecology.* New York: Oxford University Press, 1995.

Sumich, James L. *An Introduction to the Biology of Marine Life.* 5th ed. Dubuque, Iowa: William C. Brown, 1992.

POISONOUS ANIMALS

Types of ecology: Behavioral ecology; Chemical ecology; Physiological ecology

Animal poisons, or venoms, are used both as a defense mechanism and as a predatory strategy. These toxins can be delivered by biting, stinging, or body contact. Poisonous species occur throughout the animal kingdom and include snakes, insects, spiders and other arachnids, mammals, lizards, and fish.

Substances that cause disease symptoms, injure tissues, or disrupt life processes on entering the body are poisons. When ingested in large quantities, most poisons kill. Poisons can be contacted from minerals, in vegetable foods, or through animal attack. Any poison of animal origin is a venom. Venoms are delivered by biting, stinging, or other body contact. These animal poisons are used to capture prey or in self-defense. Often, it seems that the ability to make venom arose in animals that were too small, too slow, or too weak to maintain an ecological niche otherwise.

The most familiar poisonous animals are snakes, insects, spiders, and some other arachnids. Poisonous species, however, occur throughout the animal kingdom, including a few mammals and lizards, and some fish. The severity of venom effects depends on its chemical nature, the nature of the contact mechanisms, the amount of venom delivered, and victim size. For example, all spiders are poisonous. However, their venom is usually dispensed in small amounts that do not affect humans. Hence, few spiders kill humans, though they kill prey and use venom in self-defense very effectively.

Chemically, venoms vary greatly. Snake venoms are mixtures of enzymes and toxins. Study of their effects led to the identification of hemotoxins, which cause blood vessel damage and hemorrhage; neurotoxins, which paralyze nerves controlling heart action and respiration; and clotting agents, which excessively promote or prevent blood clotting. Cobras, coral snakes, and arachnids all have neurotoxic venoms.

Lizards, Arthropods, and Insects

Only two species of poisonous lizard are known: Gila monsters and beaded lizards (both holoderms). They inhabit the southwestern deserts of the United States and Mexico. They do not strike like snakes; rather, they bite, hold on, and chew to apply their venom. Holoderm bites kill prey but rarely kill humans. Beaded lizards grow to three feet long and Gila monsters grow to two feet long.

486

Most poisonous arthropods are spiders and scorpions. Both use venom to subdue or kill prey. As stated earlier, few spiders endanger humans because their venom is weak and is not injected in large quantities, but some species have very potent venom and harm or even kill humans. Best known of these are black widow spiders. Though rarely lethal to humans, black widow bites cause cramps and paralysis.

All of the approximately six hundred scorpion species, of sizes between one and ten inches, have tail-end stingers. Large, tropical scorpions can kill humans, while American scorpions are smaller and less dangerous. Scorpions are more dangerous than spiders because they crawl into shoes and other places where their habitat overlaps with that of humans.

Many insects, such as caterpillars, bees, wasps, hornets, and ants, use venom in self-defense or to paralyze prey to feed themselves or offspring. Caterpillars use poison spines for protection. Bees, wasps, hornets, and ants use stingers for the same purpose. The venom of insects also kills many organisms that seek to prey on them. Humans, however, are rarely killed by insect bites. Such bites are usually mildly to severely painful for a period from a few minutes to several days. However, for some humans who are particularly sensitive, severe anaphylaxis occurs, in some cases followed by death.

Poisonous Snakes

Poisonous snakes are colubrids, elapids, or vipers, depending on their anatomic characteristics. All have paired, hollow fangs in the front upper jaw. The fangs fold back against the upper palate when not used, and when a snake strikes they swing forward to inject a venom that attacks the victim's blood and tissues. The heads of poisonous snakes are scale-covered and triangular. Such snakes are found worldwide and include pit vipers, named for the pits on each side of the head that contain heat receptors. The pits detect warm-blooded prey, mostly rodents, in the dark. Pit vipers include rattlesnakes, moccasins, copperheads, fer-de-lance, and bushmasters.

The populations and species of American and European poisonous snakes differ. In North America, twenty such snake types occur: elapid coral snakes and copperheads, sixteen rattler types, and cottonmouths (all vipers). Vipers are found everywhere but Alaska. Rattlers have the widest habitat, as shown by their abundance in the snake-rich Great Plains, Mississippi Valley, and southern Appalachia. In contrast, copperheads and cottonmouths are abundant in Appalachia and the Mississippi Valley, respectively. Mexican poisonous snakes are divided into two ranges: the northern, from the U.S.-Mexican border to Mexico City, and the southern, south of Mexico City. In the north, snakes are mostly rattlers, as in the contiguous United States. Coral snakes and pit vipers are plentiful in the south. Most perilous are the five- to eight-foot fer-de-lance, whose venom kills many humans. All South American vipers live in tropical environments, except for rattlesnakes. Rattlers prefer arid environments, although some are also found in tropical climates. Bushmasters, the largest South American vipers, and elapid coral snakes are nocturnal and rarely endanger humans. Tropical rattlers and lance-headed vipers, somewhat less nocturnal, kill many. Europe has few snakes, due to its cool climates and scarce suitable habitats. Its few vipers range almost to the Arctic Circle. Eastern Mediterranean regions hold most of the European vipers.

There are many poisonous snakes in Africa and Asia. North Africa, mostly desert, has few snakes. Central Africa's diverse poisonous snakes are colubrid, elapid, and viper types. Elapids include dangerous black mambas, twelve to fourteen feet long, and smaller cobras, which also occur in South Africa. Among diverse vipers, the most perilous are Gaboon vipers and puff adders. The Middle East, mostly desert, has few poisonous snakes. Southeast Asia has the most poisonous snakes in the world, elapids, colubrids, and vipers. This is due to snake habitats that range from semiarid areas to rain forests. The huge human population explains why this area has the world's highest incidence of snakebite and related deaths. Vipers bite most often, but elapids cause a larger portion of deaths. The Far East snake population is complex, and its snakebite incidence is also high. Its important poisonous snakes are pit vipers.

Australia and New Guinea have large numbers of poisonous snakes. Australia has 65 percent of the world's snakes, while New Guinea has 25 percent. Also, sea snakes occur offshore and in some rivers and lakes. However, these countries have few snakebite deaths, because of the small size and nocturnal nature of most of the indigenous snakes.

Poisonous Fish and Amphibians
Venomous fish are dangerous to those who enter the oceans, especially fishermen who take them from their nets. The geographical distribution of these fish is like all other fish. The highest population density is in warm temperate or tropical waters. Numbers and varieties of poisonous fish decrease with proximity to the North and South Poles, and they are most abundant in Indo-Pacific and West Indian waters.

A well-known group of poisonous fish, the stingrays (dasyatids), inhabit warm, shallow, sandy-to-muddy ocean waters. Dasyatids lurk almost completely buried, awaiting prey that they sting to death with barbed, venomous teeth in their tails. The tail poison is made in glands at the bases of the teeth. Small, freshwater dasyatids are found in South

American rivers, such as the Amazon, hundreds of miles from the river mouths. Stingrays near Australia grow to fifteen-foot lengths. The wide distribution of stingrays and their danger to humans are mentioned in the writings of Aristotle in the third century B.C.E. and they played a role in the death of John Smith in 1608, who was killed by a stingray while exploring Chesapeake Bay.

Also well known are the venomous *Scorpaenidae* fish family, many members of which cause very painful stings. Zebrafish and stonefish are good examples. Both, like all scorpaenids, have sharp spines supporting dorsal fins. The spines, used in self-defense, have venom glands. The most deadly fish venom is that of the stonefish, which, when stepped on, can kill humans.

Frogs and Toads

Poisonous animals that endanger by contact are exemplified not only by the zebrafish and stonefish just mentioned but also by poisonous frogs or toads. Most such frogs and toads live in Africa and South America. Poison dart frogs, for example, secrete poisons through the skin. In humans, the effects of contact with these poisons range from severe irritation to death. The poisons frighten away or kill most predators that attempt to eat the frogs.

Ecological Significance

The ecological function of poisonous animals is to keep down the population of insects, rodents, arachnids, and small fish. They thus contribute to maintaining the balance of nature. Poisonous land animals, such as scorpions and many poisonous snakes, are often nocturnal and add another dimension to pest control by nighttime predation.

Sanford S. Singer

See also: Allelopathy; Defense mechanisms; Genetically modified foods; Metabolites; Pheromones; Poisonous plants; Predation.

Sources for Further Study

Aaseng, Nathan. *Poisonous Creatures*. New York: Twenty-first Century Books, 1997.

Edström, Anders. *Venomous and Poisonous Animals*. Malabar, Fla.: Krieger, 1992.

Foster, Steven, and Roger Caras. *Venomous Animals and Poisonous Plants*. Boston: Houghton Mifflin, 1994.

Grice, Gordon D. *The Red Hourglass: Lives of the Predators*. New York: Delacorte, 1998.

POISONOUS PLANTS

Types of ecology: Chemical ecology; Physiological ecology

Poisonous plants have evolved toxic substances that function to defend them against herbivores and thereby better adapt them for survival.

After evolving adaptations that facilitated colonization of terrestrial habitats, plants were confronted with a different type of problem. This was the problem of herbivory, or the inclination of many different types of organisms, from bacteria to insects to four-legged herbivores, to eat plants. Pressures from herbivory drove many different types of plants, from many different families, to evolve defenses. Some of these defenses included changes in form, such as the evolution of thorns, spikes, or thicker, tougher leaves. Other plants evolved to produce chemical compounds that make them taste bad, interrupt the growth and life cycles of the herbivores, make the herbivores sick, or kill them outright.

Phytochemicals

One of the most interesting aspects of plants, especially prevalent in the angiosperms (flowering plants), is their evolution of substances called secondary metabolites, sometimes referred to as phytochemicals. Once considered waste products, these substances include an array of chemical compounds: alkaloids, quinones, essential oils, terpenoids, glycosides (including cyanogenic, cardioactive, anthraquinone, coumarin, and saponin glycosides), flavonoids, raphides (also called oxalates, which contain needle-like crystals of calcium oxalate), resins, and phytotoxins (highly toxic protein molecules). The presence of many of these compounds can characterize whole families, or even genera, of flowering plants.

Effects on Humans

The phytochemicals listed above have a wide range of effects. In humans, some of these compounds will cause mild to severe skin irritation, or contact dermatitis; others cause mild to severe gastric distress. Some cause hallucinations or psychoactive symptoms. The ingestion of many other types of phytochemicals proves fatal. Interestingly, many of these phytochemicals also have important medical uses. The effects of the phytochemicals are dependent on dosage: At low doses, some phytochemicals are therapeutic; at higher doses, some can kill.

Alkaloids

Alkaloids are nitrogenous, bitter-tasting compounds of plant origin. More than three thousand alkaloids have been identified from about four thousand plant species. Their greatest effects are mainly on the nervous system, producing either physiological or psychological results. Plant families producing alkaloids include the *Apocynaceae, Berberidaceae, Fabaceae, Papaveraceae, Ranunuculaceae, Rubiaceae,* and *Solanaceae.* Some well-known alkaloids include caffeine, cocaine, ephedrine, morphine, nicotine, and quinine.

Glycosides

Glycosides are compounds that combine a sugar, usually glucose, with an active component. While there are many types of glycosides, some of the most important groups of potentially poisonous glycosides include the cyanogenic, cardioactive, anthraquinone, coumarin, and saponin glycosides.

Cyanogenic glycosides are found in many members of the *Rosaceae* and are found in the seeds, pits, and bark of almonds, apples, apricots, cherries, peaches, pears, and plums. When cyanogenic glycosides break down, they release a compound called hydrogen cyanide.

Two other types of glycosides, cardioactive glycosides and saponins, feature a steroid molecule as part of their chemical structure. Digitalis, a cardioactive glycoside, in the right amounts can strengthen and slow the heart rate, helping patients who suffer from congestive heart failure. Other cardioactive glycosides from plants such as milkweed and oleander are highly toxic. Saponins can cause severe irritation of the digestive system and hemolytic anemia. Anthraquinone glycosides exhibit purgative activities. Plants containing anthraquinone glycosides include rhubarb (*Rheum* species) and senna (*Cassia senna*).

Household Plants

Many common household plants are poisonous to both humans and animals. One family of popular household plants that can cause problems is the *Araceae*, the philodendron family, including plants such as philodendron and dieffenbachia. All members of this family, including these plants, contain needlelike crystals of calcium oxalate that, when ingested, cause painful burning and swelling of the lips, tongue, mouth, and throat. This burning and swelling can last for several days, making talking and even breathing difficult. *Dieffenbachia* is often referred to by the common name of dumb cane, because eating it makes people unable to talk for a few days.

Foxgloves, a common ornamental garden flower, produce cardiac glycosides with strong physiological effects on heart muscle. Although highly toxic, if processed in the right amounts these glycosides can strengthen and slow the heart rate, helping patients who suffer from congestive heart failure.
(PhotoDisc)

Landscape Plants

Many landscape plants are also poisonous. For example, the yew (genus *Taxus*), commonly planted as a landscape plant, is deadly poisonous. Children who eat the bright red aril, which contains the seed, are poisoned by the potent alkaloid taxine. Yews are poisonous to livestock as well, causing death to horses and other cattle. Death results from cardiac or respiratory failure.

Other poisonous landscape and garden plants include oleander, rhododendrons, azaleas, hyacinths, lily of the valley, daffodils, tulips, and Star-of-Bethlehem. Many legumes are also toxic, including rosary pea, lupines, and wisteria. Castor bean plant, a member of the family *Euphorbiaceae*, produces seeds that are so toxic that one seed will kill a child and three seeds are fatal to adults. The toxin produced by the seeds is called ricin, which many scientists consider to be the most potent natural toxin known.

Arrow Poisons

Toxic plant and animal products have been used for thousands of years in hunting, executions, and warfare. Usually the poisonous extracts were smeared on arrows or spears. The earliest reliable written evidence for these uses comes from the *Rigveda* from ancient India. Arrow poisons come in many different varieties, and most rain-forest hunters have their own secret blend. South American arrow poisons are generically called curare.

There are more than seventy different plant species used in making arrow poisons. Two of the main arrow poison plants are woody vines from the Amazon: *Strychnos toxifera* and *Chondodendron tomentosum*. Some types of curare have proven medically useful. They are used as muscle relaxants in surgery, which lessens the amount of general anesthetic needed. A plant called *Strychnos nux-vomica* from Asia yields the poison strychnine, a stimulant of the central nervous system.

In ancient times, toxic plant products were also commonly used in executions. Many people were expert, professional poisoners in the ancient world. They could select a poison that would take days or even months to take effect, thus ensuring, for example, that an unfaithful spouse or lover would not suspect the reason for his or her lingering illness. On occasions when a more rapid result was required, a strong dose or more powerful poison could be prescribed.

Poison Ivy
Toxicodendron radicans, commonly known as poison ivy, is well known for causing contact dermatitis. Poison ivy is a member of the *Anacardiaceae*, or cashew family, and is a widespread weed in the United States and southern Canada. It grows in a variety of habitats: wetlands, disturbed areas, and the edges of forests. It has many forms, appearing as either a shrub or a woody vine which will grow up trees, houses, fences, and fence posts. It has alternate leaves with three leaflets, forming the basis of the old saying "Leaves of three, let it be." After poison ivy flowers, it develops clusters of white or yellowish-white berries. Related species are poison oak, western poison oak, and poison sumac, which some scientists consider to be different types of poison ivy.

Roughly half the world's population is allergic to poison ivy. Very sensitive people develop a severe skin rash; about 10 percent of the people who are allergic require medical attention after exposure. The chemical compound causing the allergic reaction is called *urushiol*, a resin found in all parts of the plant. Urushiol is so potent that in some individuals, just one drop produces a reaction. Inhaling smoke from burning poison ivy can result in eye and lung damage. For some people, mere contact with the smoke from burning poison ivy can trigger a reaction. Urushiol lasts forever; in herbaria, dried plants one hundred years old have given unlucky botanists contact dermatitis.

Carol S. Radford

See also: Allelopathy; Defense mechanisms; Genetically modified foods; Metabolites; Pheromones; Poisonous animals; Predation.

Sources for Further Study

Burrows, George E., and Ronald J. Tyrl. *Toxic Plants of North America*. Ames: Iowa State University Press, 2001.

Levetin, Estelle, and Karen McMahon. *Plants and Society*. 2d ed. Boston: WCB/McGraw-Hill, 1999.

Lewis, Walter H., and Memory P. F. Elvins-Lewis. *Medical Botany: Plants Affecting Man's Health*. New York: John Wiley and Sons, 1997.

Simpson, Beryl B., and Molly Conner Ogarzaly. *Economic Botany: Plants in Our World*. 3d ed. Boston: McGraw-Hill, 2000.

POLLINATION

Types of ecology: Community ecology; Physiological ecology

Pollination—the transfer of pollen from anther to stigma in flowering plants or from male cone to ovules in gymnosperms—accounts for a wide variety of ecological interactions in communities of organisms.

Pollination is the process, in sexually reproducing plants (both angiosperms and gymnosperms), whereby the male sperm and female egg are joined via transfer of pollen (male microspore). If the anthers and stigmas of the plants involved have the same genetic makeup or they are produced on the same plant, the type of pollination is called self-pollination. If anthers and stigmas are from plants with different genetic makeups, the type of pollination is called cross-pollination.

Self-pollination is efficient because pollen from the anther of a flower can be transferred easily onto the stigma of the same flower, owing to the proximity of the two parts. On the other hand, cross-pollination is risky because the transfer of pollen involves long distances and precise destinations, both of which depend on animal pollinators. In areas with few animal pollinators, the opportunities for cross-pollination may be greatly reduced (one of the many reasons that preserving biological diversity is an important ecological issue).

In spite of the risk associated with cross-pollination, most flowers have mechanisms that promote this kind of pollination. Cross-pollination increases the likelihood that offspring are vigorous, healthy, fertile, and able to survive even if the environment changes. Self-pollination leads to offspring that are less vigorous, less productive, and more subject to inbreeding depression (weakening of the offspring as a result of inbreeding).

When certain consumers forage among plants for food, they often come in contact with flowers. Many insects and other animals become dusted with pollen, and in the course of their travel they unintentionally but effectively bring about pollination. Throughout the evolutionary history of flowering plants, many pollinators have coevolved with plants. Coevolution occurs when the floral parts of a plant and the body parts and behavior of the pollinators become mutually adapted to each other, thereby increasing the effectiveness of their interaction. In many instances, the relationship between the plant and pollinator has become highly specialized, resulting in mutualism, which is interaction where both organisms benefit from each other.

In the case of pollination by animals, the pollinator receives a reward from the flower in the form of food. When the pollinator moves on, the plant's pollen is transferred to another plant. The adaptations between the flower and its pollinators can be intricate and precise and may even involve force, drugs, deception, or sexual enticement. In flowering plants, pollination is mostly due to insects or wind, but birds, bats, and rodents also act as pollinators for a number of plants.

Insects

Insect pollination occurs in the majority of flowering plants. There is no single set of characteristics for insect-pollinated flowers, because insects are a large and diverse group of animals. Rather, each plant may have a set of reproductive features that attracts mostly a specific species of insect. The principal pollinating insects are bees, although many other kinds of insects act as pollinators, including wasps, flies, moths, butterflies, ants, and beetles.

Bees have body parts suitable for collecting and carrying nectar and pollen. Their chief source of nourishment is nectar, but they also collect pollen for their larvae. The flowers that bees visit are generally brightly colored and predominantly blue or yellow—rarely pure red, because red appears black to bees. The flowers they visit often have distinctive markings that function as guides that lead them to the nectar. Bees can perceive

Honeybees are well known for their contribution to plant propagation, carrying pollen from flower to flower. This relationship has coevolved over time; the bees are dependent on the flowers, and vice versa. (PhotoDisc)

ultraviolet (UV) light (a part of the spectrum not visible to humans), and some flower markings are visible only in UV light, making patterns perceived by bees sometimes different from those seen by humans. Many bee-pollinated flowers are delicately sweet and fragrant.

Moth- and butterfly-pollinated flowers are similar to bee-pollinated flowers in that they frequently have sweet fragrances. Some butterflies can detect red colors, and so red flowers are sometimes pollinated by them. Many moths forage only at night; the flowers they visit are usually white or cream-colored because these colors stand out against dark backgrounds in starlight or moonlight. With their long mouthparts, moths and butterflies are well adapted for securing nectar from flowers with long, tube-shaped corollas (the petals collectively), such as larkspur, nasturtium, tobacco, evening primrose, and amaryllis.

The flowers pollinated by beetles tend to have strong, yeasty, spicy, or fruity odors. They are typically white or dull in color, in keeping with the diminished visual sense of their pollinators. Although some beetle-pollinated flowers do not secrete nectar, they furnish pollen or other foods which are available on the petals in special storage cells.

Birds

Birds and the flowers that they pollinate are also adapted to each other. Birds do not have a highly developed sense of smell, but they have a keen sense of vision. Their flowers are thus frequently bright red or yellow and usually have little, if any, odor. The flowers are typically large or are part of a large inflorescence. Birds are highly active pollinators and tend to use up their energy very rapidly. Therefore, they must feed frequently to sustain themselves. Many of the flowers they visit produce copious quantities of nectar, assuring the birds' continued visitation. The nectar is frequently produced in long floral tubes, which prevent most insects from gaining access to it. Examples of bird-pollinated flowers are red columbine, fuchsia, scarlet passion flower, eucalyptus, hibiscus, and poinsettia.

Bats and Rodents

Bat-pollinated flowers are found primarily in the tropics, and they open only at night, when the bats are foraging. These flowers are dull in color, and like bird-pollinated flowers, they are large enough for the pollinator to insert part of its head inside. The plants may also consist of ball-like inflorescences containing large numbers of small flowers whose stamens readily dust the visitor with pollen. Bat-pollinated flowers include bananas, mangoes, kapok, and sisal. Like moth-pollinated flowers, flowers that attract bats and small rodents open at night. Mammal-pollinated flowers are usually white and strongly scented, often with a fruity odor. Such flowers are large, to provide the pollinators enough pollen and nectar to fulfill their energy requirements. The flowers are also sturdy, to bear the frequent and vigorous visits of these small mammals.

Orchid Pollinators

The orchid family has pollinators among bees, moths and butterflies, and beetles. Some of the adaptations between orchid flowers and their pollinators are extraordinary. Many orchids produce their pollen in little sacs called pollinia, which typically have sticky pads at the bases. When a bee visits such a flower, the pollinia are usually deposited on its head. In some orchids, the pollinia are forcibly "slapped" on the pollinator through a trigger mechanism within the flower. In some orchids, a petal is modified so that it resembles a female wasp or bee. Male wasps or bees emerge from their pupal stage before the females and can mistake the orchids for potential mates. They try to copulate with these flowers, and while they are doing so, pollinia are deposited on their heads. When the wasps or bees visit other flowers, the pollinia are caught in sticky stigma cavities.

When moths and butterflies pollinate orchids, the pollinia become attached to their long tongues by means of sticky clamps instead of pads. The pollinia of certain bog orchids become attached to the eyes of the female mosquitoes that pollinate them. After a few visits, the mosquitoes are blinded and unable to continue their normal activities (a good example of a biological control within an ecosystem).

Among the most bizarre of the orchid pollination mechanisms are those whose effects are to dunk the pollinator in a pool of watery fluid secreted by the orchid itself and then permit the pollinator to escape underwater through a trap door. The route of the insect ensures contact between the pollinia and stigma surfaces. In other orchids with powerful narcotic fragrances, pollinia are slowly attached to the drugged pollinator. When the transfer of pollinia has been completed, the fragrance abruptly fades away, and the insect recovers and flies away.

Wind and Water

Wind pollination is common in those plants with inconspicuous flowers, such as grasses, poplars, walnuts, alders, birches, oaks, and ragweeds. These plants lack odor and nectar and are, hence, unattractive to insects. Furthermore, the petals are either small or absent, and the sex organs are often separate on the same plant. In grasses, the stigmas are feathery and expose a large surface to catch pollen, which is lightweight, dry, and easily blown by the wind. Because wind-pollinated flowers do not depend on animals to transport their pollen, they do not invest in the production of rewards for their visitors. However, they have to produce enormous quantities of pollen. Wind pollination is not efficient because most of the pollen does not end up on the stigmas of appropriate plants but on the ground, bodies of water, and in people's noses (a major cause of allergic reactions). Wind pollination is successful in cases where a large number of individuals of the same species grow fairly close together, as in grasslands and coniferous forests.

Water pollination is rare, simply because fewer plants have flowers that are submerged in water. Such plants include the sea grasses, which release pollen that is carried passively by water currents. In some plants, such as the sea-nymph, pollen is threadlike, thus increasing its chances of coming in contact with stigmas. In eelgrass, the entire male flower floats.

Danilo D. Fernando

See also: Adaptations and their mechanisms; Animal-plant interactions; Coevolution; Communities: ecosystem interactions; Communities: structure; Reproductive strategies; Symbiosis.

Sources for Further Study

Barth, Friedrich G. *Insects and Flowers: The Biology of a Partnership*. Princeton, N.J.: Princeton University Press, 1991.

Proctor, Michael, Peter Yeo, and Andrew Lack. *The Natural History of Pollination*. Portland, Oreg.: Timber Press, 1996.

Raven, Peter H., Ray F. Evert, and Susan E. Eichhorn. *Biology of Plants*. 6th ed. New York: W. H. Freeman/Worth, 1999.

POLLUTION EFFECTS

Type of ecology: Ecotoxicology

Pollutants in soil, water, and atmosphere have created enormous problems for the living world. Destroyed habitats and polluted food sources and drinking water for animals have caused deformations in animal growth, development, and reproduction, as well as a shortening of life span, all of which contribute to an accelerated decrease in biodiversity and the extinction of more species.

During the last decade of the twentieth century, the ecological problems predicted by environmental scientists decades previously began to accelerate in a variety of ways. These included the human population explosion, food imbalances, inflation brought about by energy resource scarcity, acid rain, toxic and hazardous wastes, water shortages, major soil erosion, a punctuated ozone layer, and greenhouse effects. As a result of pollution, decreases in biodiversity and the extinction of both plant and animal species has accelerated. The burning and cutting of thousands of square miles of rain forests not only destroyed habitats for numerous animal species but also caused irreversible damage to ecosystems and climates. Industrialization and the expansion of the human population had left relatively few places on earth undisturbed. Heavy dependence upon fossil fuels for energy and synthetic chemicals has resulted in the dumping of millions of metric tons of nonnatural compounds and chemicals into the environment.

Recurrent drought and famine in Africa testify to human mischief toward Mother Nature. The well-being of animals as well as humans will not be protected against the ecological consequences of human actions by remaining ignorant of those actions. Effective measures taken to reduce pollution and protect natural resources and the environment first come with a recognition of these problems. The ignorance and inaction of ordinary citizens will lead to disastrous consequences for the environment, threatening humanity's very existence.

Sources and Types of Pollution

Among the primary sources of pollution are agrichemicals such as fertilizers, insecticides, fungicides, and herbicides. The application of excess chemical fertilizers applied to soil hampers natural cycling of nutrients, depletes the soil's own fertility, and destroys the habitats of thousands of small animals residing in the soil. Farm runoff carries priceless topsoil, expensive fertilizer, and animal manure into rivers and lakes, where these potential resources become pollutants and cause eutrophication and the subsequent death of fish and other wildlife.

In the city, water pours from sidewalks, rooftops, and streets, picking up soot, silt, oil, heavy metals, and garbage. It races down gutters into storm sewers, carrying household pollutants from cleaning solutions to prescription medications, and a weakly toxic soup gushes into the nearest stream, river, or ocean. Many of these chemicals also seep into the ground, causing contamination of groundwater.

Plants and factories manufacturing these chemical products are another source of pollutants and contamination. Burning fossil fuels releases greenhouse gases, carbon dioxide, and methane. Coupled with deforestation in many regions of the world, carbon dioxide concentration in the atmosphere has steadily climbed, from 290 parts per million in 1860 to 370 parts per million in 1990, a more than 25 percent rise due to industrialization. The resultant global warming will have far-reaching effects on plants, animals, and humans in ways still not understood. Acid rain, a result of overcharging the atmosphere with nitric oxides and sulfur dioxide (two gases also released by burning of fossil fuels), has increased the acidity of soil and lakes to levels at which many organisms cannot survive. The most acidic rain is concentrated in the Northeast of the United States. In New York's Adirondack Mountains, for instance, acid rain has made about a third of all the lakes and ponds too acidic to support fish. First, much of the food web that sustains the fish was destroyed. Clams, snails, crayfish, and insect larvae die first, then amphibians, and finally fish. The detrimental effect is not limited to aquatic animals. The loss of insects and their larvae and small aquatic animals has contributed to a dramatic decline in the population of black ducks that feed on them. The result is a crystal-clear lake, beautiful but dead.

Another serious problem created by the chemical industry is ozone depletion. Chlorofluorocarbon (CFC) compounds contain chlorine, fluorine, and carbon. Since their development in the 1930's, these compounds were widely used as coolants in refrigerators and air conditioners, as aerosol spray propellants, as agents for producing Styrofoam, and as cleansers for electronic parts. These chemicals are very stable and for decades were considered to be safe. Their stability, however, turned out to be a real problem. They were in gaseous form and rose into the atmosphere. There, the high energy level of ultraviolet (UV) light breaks them down, releasing chlorine atoms, which in turn catalyzes the breakdown of ozone to oxygen gas. As a result of the decline of ozone and the punctuation of the ozone layer, UV radiation has risen by an average of 8 percent per decade since the 1970's.

Air Pollutant Emissions by Pollutant and Source, 1998

Source	Particulates[1]	Sulfur Dioxide	Nitrogen Oxides	Volatile Organic Compounds	Carbon Monoxide	Lead (tons)
Fuel Combustion (stationary sources)						
Electric utilities	302	13,217	6,103	54	417	68
Industrial	245	2,895	2,969	161	1,114	19
Other fuel combustion	544	609	1,117	678	3,843	416
Residential	432	127	742	654	3,699	6
Subtotal	1,091	16,721	10,189	893	5,374	503
Industrial processes						
Chemical and allied product manufacturing	65	299	152	396	1,129	175
Metals processing	171	444	88	75	1,495	2,098
Petroleum and related industries	32	345	138	496	368	NA
Other	339	370	408	450	632	54
Subtotal	607	1,458	786	1,417	3,624	2,327
Solvent utilization	6	1	2	5,278	2	NA
Storage and transport	94	3	7	1,324	80	NA
Waste disposal and recycling	310	42	97	433	1,154	620
Highway vehicles						
Light-duty gas vehicles and motorcycles	56	130	2,849	2,832	27,039	12
Light-duty trucks	40	99	1,917	2,015	18,726	7
Heavy-duty gas vehicles	8	11	323	257	3,067	—
Diesels	152	86	2,676	222	1,554	NA
Subtotal	257	326	7,765	5,325	50,386	19
Off highway[2]	461	1,084	5,280	2,461	19,914	503
Miscellaneous[3]	31,916	12	328	786	8,920	NA
Total emissions	34,742	19,647	24,454	17,917	89,454	3,972

Source: Adapted from U.S. Environmental Protection Agency, *National Air Pollutant Emission Trends*, 1900-1998, EPA-454/R-00-002. From *Statistical Abstract of the United States: 2001* (Washington, D.C.: U.S. Bureau of the Census, 2001).

Note: In thousands of tons, except as indicated.

— Represents or round to zero.

NA Not available

[1] Represents both particulates of less than 10 microns and particulate dust from sources such as agricultural tilling, construction, mining, and quarrying, paved roads, unpaved roads, and wind erosion.

[2] Includes emissions from farm tractors and other farm machinery, construction equipment, industrial machinery, recreational marine vessels, and small general utility engines such as lawn mowers.

[3] Includes emissions such as from forest fires and other kinds of burning, various agricultural activities, fugitive dust from paved and unpaved roads, and other construction and mining activities, and natural sources.

This depletion of the ozone layer poses a threat to humans, animals, plants, and even microorganisms.

Long-Range Impacts of Pollution
The degradation of air, land, and water as a result of the release of chemical and biological wastes has wide-ranging effects on animals. On a large scale, pollution destroys habitats and produces population crashes and even the extinction of species. Hazardous chemicals introduced into the environment sometimes render an environment unfit for life (as at Love Canal, New York, or Times Beach, Missouri). At the individual level, pollution causes abnormalities in growth, development, and reproduction. Hazardous chemicals, introduced either intentionally (such as fertilizers, herbicides, and pesticides) or through neglect (as with industrial wastes), have a variety of detrimental, sometimes devastating effects on animals. They affect the metabolism, growth and development, reproduction, and average life spans of many species.

A few examples will illustrate the effects of chemical pollution on animals. In the 1940's, the new insecticide dichloro-diphenyl-trichloroethane (DDT) was regarded as a miracle. It saved millions of lives in the tropics by killing the mosquitoes that spread deadly malaria. DDT saved millions more lives with increased crop yields resulting from DDT's destruction of insect pests. This miraculous pesticide, however, turned out to be a long-lasting nemesis to many species of wildlife and the environment. In the United States, ecologists and wildlife biologists during the 1950's and 1960's witnessed a stunning decline in the populations of several predatory birds, especially fish-eaters, such as bald eagles, cormorants, ospreys, and brown pelicans. The population decline drove the brown pelican and bald eagle close to extinction. In 1973, the U.S. Congress passed the Endangered Species Act, which banned the use of DDT. The once-threatened species have somewhat recovered since. In the mid 1950's, the World Health Organization used DDT on the island of Borneo to control malaria. DDT entered food webs through a caterpillar. Wasps that fed on caterpillars were first destroyed. Gecko lizards that ate the poisoned insects accumulated high levels of DDT in their bodies. Both geckos and the village cats that ate the geckos died of DDT poisoning. The rat population exploded with its natural enemy, cats, eliminated. The village was then threatened with an outbreak of plague, carried by the uncontrolled rats.

Although DDT has been banned in much of the world, there is a growing concern over the effects of a number of chlorinated compounds. These chemicals, described as "environmental estrogens," interfere with normal sex hormone functions by mimicking the effects of the hormone estrogen or enhancing estrogen's potency. High levels of chlorinated compounds, such as dioxin and polychlorinated biphenyls (PCBs), in the Great Lakes have led to a sharp decline in populations of river otters and a variety of fish-eating birds, including the newly returned bald eagles. These chemicals are also the cause of deformed offspring, or eggs that never hatch. In Florida's Lake Apopka, a spill of chlorinated chemicals in 1980 led to a 90 percent drop in the birthrate of the lake's alligators. These are only a few examples of the detrimental effects on various animals by synthetic chemicals.

Air Pollution

Air pollution leads to acid rain and the greenhouse effect, as well as damage to the ozone layer. Acid rain drops out of the skies onto areas at great distances from the source of the acids and destroys forests and lakes in sensitive regions. As a result, fish populations are dwindling or being eliminated in lakes and streams by a lower pH caused by acid deposition. The strongest evidence comes from data collected from the past twenty-five to forty-five years in Adirondack lakes and in Nova Scotia rivers. Studies during this period clearly show declines in acid-sensitive species. Similar results were obtained from analyzing fish population and water acidity in Maine, Massachusetts, Pennsylvania, and Vermont. The consensus is that fish populations would be eliminated if the surface waters acidify to between pH 5.0 to pH 5.5. The effects of acid rain on other animals are indirect, either through the dwindling fish population (as a food source for other animals) or stunted forest growth (disturbance to habitats).

The effect of global warming on the animal kingdom is also a serious and complex issue. As global temperature rises, ice caps in polar regions and glaciers melt, ocean waters expand in response to atmospheric warming, and thus the sea level elevates. The expected sea level rise will flood coastal cities and coastal wetlands. These threatened ecosystems are habitats and breeding grounds for numerous species of birds, fish, shrimp, and crabs, whose populations could be severely diminished. The Florida Everglades will virtually disappear if the sea level rises two feet. The impact of global warming on forests could be profound. The distribution of tree species is exquisitely sensitive to average annual temperature, and small changes could dramatically alter the extent and species composition of forests. This in turn could dramatically alter the population distribution of animals, and hence biodiversity.

The effect of the punctuated ozone layer on animals is yet to be fully understood. It is known that the high energy level of UV radiation can damage biological molecules, including the genetic material deoxyribonucleic acid (DNA), causing mutation. In small quantities, UV light helps the skin of humans and many animals produce vitamin D and causes tanning. However, in large doses, UV causes sunburn and premature aging of skin, skin cancer, and cataracts, a condition in which the lens of the eye becomes cloudy. Due to UV radiation's ability to penetrate, even animals covered by hair and thick fur cannot escape from these detrimental effects. Ozone damage costs U.S. farmers over $2 billion annually in reduced crop yields. All who depend on forestry and agriculture may bear a much higher cost if the emission of pollutants that destroy ozone are not regulated soon.

Possible Remedies

The various types of pollution all have serious effects on the plant and animal species that share this planet. It is all too easy to document the impacts of pollution on human health and ignore their effects on the rest of the living world. Any possible remedies to alleviate these problems should start with education, the realization of these problems at an individual as well as a global level. The tasks seem to be insurmountable, and no organization, no country can do it alone. It takes willingness to accept short-term inconvenience or economic sacrifice for long-term benefit. A couple of examples serve to illustrate what can be done to alleviate the problems of pollution.

Synthetic chemical pollutants that are poisoning both people and wildlife could be largely eliminated without disrupting the economy, as reported in a study published in 2000 by the Worldwatch Institute, a Washington, D.C.-based environmental organization. The report presents strong evidence from three sectors that are major sources of these pollutants—paper manufacturing, pesticides, and PVC plastics—to show that nontoxic options are available at competitive prices. Agricultural pollution can be mitigated, significantly reduced, or virtually eliminated through the use of proper regulation and economic incentives. Farmers from Indonesia to Kenya are learning how to use less of various chemicals while boosting yields. Since 1998, all farmers in China's Yunnan Province have eliminated their use of fungicides, while doubling rice yields, by planting more diverse varieties of the grain. In most, if not all, cases, the question is not whether it is possible to alleviate the pollution of the environment; rather, it is whether we realize the urgency and/or are willing to take a high road to do it. For the common well-being of generations to come, better approaches have to be taken to preserve the environment and biodiversity.

Ming Y. Zheng

Pollution effects

See also: Acid deposition; Biological invasions; Biomagnification; Biopesticides; Deforestation; Eutrophication; Genetically modified foods; Integrated pest management; Invasive plants; Ocean pollution and oil spills; Ozone depletion and ozone holes; Pesticides; Phytoplankton; Slash-and-burn agriculture; Waste management.

Sources for Further Study

Brown, Lester. *State of the World 2000*. New York: W. W. Norton, 2000.

Hill, Julia B. *The Legacy of Luna: The Story of a Tree, a Woman, and the Struggle to Save the Redwoods*. San Francisco: HarperSanFrancisco, 2000.

Johnson, Arthur H. "Acid Deposition: Trends, Relationships, and Effects." *Environment* 28, no. 4 (May, 1986): 6-11, 34-39.

Lippmann, Morton, ed. *Environmental Toxicants: Human Exposures and Their Health Effects*. 2d ed. New York: John Wiley & Sons, 2000.

Sampat, Payal. *Deep Trouble: The Hidden Threat of Groundwater Pollution*. Washington, D.C.: Worldwatch Institute, 2000.

POPULATION ANALYSIS

Type of ecology: Population ecology

Many animal populations are becoming threatened or endangered, primarily due to loss of suitable habitat. Population analysis enables biologists to examine the factors which lead to declines in animal populations and thus is important in the management of wild species.

A population is a group of organisms belonging to the same species that occur together in the same time and place. Population analysis is the study of biological populations, with the specific intent of understanding which factors are most important in determining population size. Populations can change over time. They increase or decrease in size, and their change in size can depend on a wide variety of factors. For example, a wildlife biologist might be interested in studying the population of porcupines that inhabits a hemlock forest or the population of bark beetles that lives on a particular tree.

Population analysis is the study of biological populations, with the specific intent of understanding which factors are most important in determining population size. Factors such as the per capita rates of birth and death, the population density, age structure, and sex ratio all contribute to determine population size. Understanding how these factors interact to influence population size is critical if biologists hope to manage populations of organisms at sustainable levels for hunting or fishing and if conservation biologists hope to prevent populations from going extinct.

Discrete vs. Continuous Populations

In order to conduct a population analysis, one must first determine whether the population of interest is best understood as discrete or continuous. A discrete population is one in which important events such as birth and death happen during specific intervals of time. A continuous population is one in which births, deaths, and other events take place continuously through time. Many discrete populations are those with nonoverlapping generations. For example, in many insect populations, the adults mate and lay eggs, after which the adults die. When the juveniles achieve adulthood, their parental generation is no longer living. In contrast, most continuous populations also have overlapping generations. For instance, in antelope jackrabbits (*Lepus alleni*), females may give birth at any time during the year, and members of several generations occur together in space and time.

Mathematical Models

The dynamics of animal populations are affected by a wide variety of demographic factors, including the population birthrate, death rate, sex ratio, age structure, and rates of immigration and emigration. In order to understand the effects of these factors on a population, biologists use population models. A model is an abstract representation of a concrete idea. The representation created by the model boils the concrete idea down into a few critical components. By building and examining population models, population analysts investigate the relative importance of different factors to the dynamics of a given population.

A basic mathematical model of population size is as follows:

$$N_{t+1} - N_t + B - D + I - E \qquad \text{(equation 1)}$$

where N_t+1 equals the population size after one time interval, N_t equals the total number of individuals in the population at the initial time, B equals the number of births, D equals the number of deaths, I equals the number of immigrants into the population, and E equals the number of emigrants leaving the population. This simple model boils population size down to just four factors, B, D, I, and E. This model is not meant to be a true or precise representation of the population; rather, it is meant to clarify the importance of the factors of birth, death, immigration, and emigration on population size. To use the same model to examine the rate of growth of a population through time, it can be rearranged as follows:

$$N_{t+1} - N_t = B - D + I - E \qquad \text{(equation 2)}$$

That is, the increase or decrease in the population size between time intervals t and $t+1$ is reflected by the number of births, deaths, immigrants, and emigrants.

When population biologists choose to focus specifically on the importance of birth and death in population dynamics, population models are simplified by temporarily ignoring the effects of immigration and emigration. In this case, the degree of change in the population between time intervals t and $t+1$ becomes:

$$N_{t+1} - N_t = B - D \qquad \text{(equation 3)}$$

It is usually safe to assume that the total number of births (B) and deaths (D) in a population is a function of the total number of individuals in the population at the time, N_t. For example, if there are only ten females in a population at time t, it would be impossible to have more than ten births in the population. More births and deaths are possible in larger populations. If B equals the total number of births in the population, then B is equal to the rate at which each individual in the population gives birth, times the total number of individuals in the population. Likewise, the total number of deaths, D, will be equal to the rate at which each individual in the population might die times the total number of individuals in the population. In other words:

$$B = bN_t \text{ and } D = dN_t \qquad \text{(equation 4)}$$

where b and d represent the per capita rate of birth and death, respectively. Given this understanding of B and D, the original model becomes

$$N_{t+1} - N_t = (bN_t) - (dN_t)$$

or

$$N_{t+1} - N_t = (b - d)N_t \qquad \text{(equation 5)}$$

It would be useful to find a variable that can represent per capita births and deaths at the same time. Biologists define r as the per capita rate of increase in a population, which is equal to the difference between per capita births and per capita deaths:

$$r = b - d \qquad \text{(equation 6)}$$

Thus, the equation that examines the changes in population size between time intervals t and $t + 1$ becomes:

$$N_{t+1} - N_t = rN_t \qquad \text{(equation 7)}$$

A numerical example works as follows. In a population that originally has 1,000 individuals, a per capita birthrate of 0.1 birth per year and a per capita death rate of 0.04 death per year, the net change in the population size between the year t and $t + 1$ would be:

$$r = 0.1 - 0.04 = 0.06$$

$$N_{t+1} - N_t = 0.06(1000) = 60$$

In other words, the population would increase by sixty individuals over the course of one year.

Continuous Populations

This model works for populations in which events take place during discrete units of time, such as a population of squirrels in which reproduction takes place at only two specific times in a single year. In contrast, many populations are continuously reproductive. That is, at any given time, any female in the population is capable of reproducing. When these conditions are met, time is viewed as more fluid than discrete, and the population exhibits continuous growth. Models of population growth are slightly different when births and deaths are continuous rather than discrete. One way to imagine the difference between a population with continuous rather than discrete growth is to imagine a population in which each time interval is infinitesimally small. When these conditions are met, the model for population growth becomes:

$$\delta N / \delta t = rN \qquad \text{(equation 8)}$$

where $\delta N / \delta t$ represents the changes in numbers in the population over very short time intervals. The per capita rate of increase (r) can now also be called the instantaneous rate of increase because the population is one with minute time intervals.

Choosing the Right Model

How does a population biologist select the best model? Which model is best depends on exactly what it is that a scientist is trying to understand about a population. In the first model presented above (equation 1), the different effects of birth, death, immigration and emigration can be compared relative to one another. In the second model, the effects of immigration and emigration are ignored and the effects of birth and death are summarized into one constant called the per capita rate of increase (equations 7 and 8). If the scientist is trying to understand the cumulative effects of B, D, I, and E on the population, then equation 1 would represent a good model. On the other hand, if the scientist is trying to understand how births and deaths influence the net changes in population size, equation 7 or 8 would be a better model.

When dealing with a continuous rather than a discrete population, equation 8 represents the rate of population growth as a function of per capita births and per capita deaths in the population. Equation 8 represents a population that is growing exponentially without bound. In other words, regardless of the population size at any given time, the per capita rate of increase remains the same. It would be reasonable to assume that per capita rates of increase can actually change with changes in overall population size. For example, in a population of bark beetles inhabiting the trunk of a tree, many more resources are available to individual beetles when the population is small. Resources must be shared between more and more individuals as the population size increases, which can result in changes to the per capita rate of increase. A model of population growth that incorporates the effect of overall population density on the per capita rate of increase might look like this:

$$\delta N / \delta t = r(1 - N/K)N \qquad \text{(equation 9)}$$

where K is equal to the carrying capacity, the maximum number of individuals in the population that there are adequate resources to support. The per capita rate of increase in equation 9 is not simply r by itself, but becomes $r(1 - N/K)$. The per capita rate of increase is a function of rates of birth and death scaled by the population size and the carrying capacity of the habitat. If the population is very large relative to the number of individuals that the habitat can support, then $N \approx K$, and the expression $(1 - N/K)$ becomes approximately equal to 0. When so, equation 9 takes the following form:

$$\delta N / \delta t = r(0)N = 0 \qquad \text{(equation 10)}$$

and the rate of population growth is zero. In other words, the population has ceased growing. On the other hand, if the population is very small relative to the number of individuals the habitat can support, then $N \ll K$ and the expression $(1 - N/K)$ becomes approximately equal to 1. When so, equation 9 takes the form

$$\delta N / \delta t = r(1)N = rN \qquad \text{(equation 11)}$$

and the rate of population growth remains a function of the rates of birth and death, but not the population size or carrying capacity. Thus, equation 9 represents what is called density-dependent growth.

Sex Ratio and Age Structure

The model set forth in equation 9 takes into account only those ways in which births, deaths, and population density relative to carrying capacity influence population growth. Sometimes it is helpful to understand how other factors such as the sex ratio and age structure in a population influence rates of growth. For example, deer hunters are not always allowed to take equal numbers of bucks and does from a population. Similarly, fishermen are often restricted in the size of fish they are allowed to keep when fishing. These wildlife management and population analysis restrictions on the sex and size of animals that can be hunted arise from the fact that both age and sex can influence population growth rates. Models that incorporate the effects of age structure and population sex ratios will not be covered here. Suffice it to say that a population that consists mostly of young individuals yet to reproduce will grow more quickly than an equally sized population of mostly older individuals who have finished reproducing. Similarly, a population with a highly skewed sex ratio that has many more males than females will not grow as quickly as a population of equal size in which the number of males and females is equal.

Erika L. Barthelmess

See also: Biodiversity; Biogeography; Clines, hybrid zones, and introgression; Demographics; Extinctions and evolutionary explosions; Gene flow; Genetic diversity; Genetic drift; Human population growth; Nonrandom mating, genetic drift, and mutation; Population fluctuations; Population genetics; Population growth; Reproductive strategies; Speciation; Species loss.

Sources for Further Study

Gardali, Thomas, et al. "Demography of a Declining Population of Warbling Vireos in Coastal California." *The Condor* 102, no. 3 (August, 2000): 601-609.

Hastings, Alan. *Population Biology: Concepts and Models.* New York: Springer-Verlag, 1997.

Hedrick, Philip W. *Population Biology: The Evolution and Ecology of Populations.* Boston: Jones and Bartlett, 1984.

Johnson, Douglas H. "Population Analysis." *Research and Management Techniques for Wildlife and Habitats*, edited by Theodore A. Bookhout. 5th ed. Bethesda, Md.: The Wildlife Society, 1994.

POPULATION FLUCTUATIONS

Type of ecology: Population ecology

The simplest realistic models of population growth produce populations that rise to some level and then stay there. These models cannot produce the complicated array of fluctuations observed in natural populations. Fluctuations vary in period from a few weeks to many decades and can reach sufficient amplitude to threaten populations (and entire species) with extinction.

The number of organisms making up a population is never constant; it always changes over time. The populations of some species change in predictable or cyclical ways, whereas populations of other species frequently exhibit seemingly unpredictable and noncyclic changes. Fluctuations in population size may be caused by changes in the population's environment; for example, seasonal changes in temperature or moisture produce seasonal fluctuations in population size. Resource limitations may produce density-dependent reductions in the growth rate of a population, which, if the reduction is not instantaneous, can result in oscillations in population size. Interactions with other species also produce population fluctuations; mathematical models of predator-prey systems typically produce oscillations in the abundance of both predators and prey. Finally, natural or anthropogenic disturbances often reduce the size of a population, which then either recovers its former abundance over time or declines further to local (or global) extinction.

The Time Scales of Population Fluctuations
Population fluctuations occur over many different time scales. On a geologic time scale (occurring over millions of years), species arise, increase to some level of abundance, and finally become extinct. These long-term patterns of species abundance provide a background for understanding population fluctuations that occur over ecological time (over days, weeks, years, or centuries). Fluctuations on these briefer time scales draw most of the attention of ecologists interested in population dynamics.

Many species of animals, including numerous insects and several small vertebrates, exhibit a more or less annual life cycle, characterized by increasing numbers and higher levels of activity during the summer (or wet season) and by dormancy or decreasing numbers during the winter (or dry season). Even highly mobile animals, such as birds, exhibit a strong seasonal pattern of abundance, if viewed from a local perspective; in North

America, for example, most songbirds migrate to more tropical latitudes in the fall and to temperate latitudes in the spring, thereby producing a yearly cycle of abundance in each location. Yearly cycles of abundance are predictable and easily explainable in terms of seasonal patterns of temperature, moisture, and sunlight. Of more interest to ecologists are population fluctuations that appear to be random or unpredictable from year to year or those fluctuations that occur out of synchrony with climatic cycles.

Regular Fluctuations

Nonseasonal fluctuations are of two main types: those that exhibit more or less regular cycles of abundance over several years and those that seem to fluctuate irregularly or noncyclically. A three- to four-year cycle of abundance is characteristic of several species of mice, voles, and other rodents found in far northern latitudes. Probably the best-known example of this type of cycle is that observed in lemming species in the northern tundra of Europe and North America. Lemming populations exhibit very high densities every three to four years, with such low densities in the intervening years that they are difficult to locate and study. This boom-or-bust cycle is apparently caused by alternating selection regimes. When lemmings are rare, high reproductive capacity and nonaggressive social behavior are favored, and the population grows rapidly. As the growing population becomes more crowded, aggressive individuals are favored, because they can hold territories, secure mates, and protect offspring better than passive individuals. The aggressive interactions, however, inhibit reproductive capacity, increase mortality attributable to fighting and infanticide, and expose more lemmings to predation as subordinate individuals are forced by dominants to occupy more marginal habitats. The behavioral changes that occur in response to crowding apparently persist for some time even as the density declines, so that aggressive interactions and a depressed birthrate continue until the lemming population reaches very low levels. Finally, passive individuals with high reproductive rates are again favored, and the cycle repeats.

Although the breeding cycles of many predators, including snowy owls, weasels, and foxes, are tied to lemming abundance, it appears that the regular fluctuation of lemming populations is a product of crowding and resource limitation rather than of a classical predator-prey cycle; that is, there is no tight coupling between the population fluctuations of lemmings and those of their predators. There is, however, a tight coupling between the population cycles of the snowshoe hare and the Canadian lynx. Beginning in 1800, the Hudson's Bay Company kept records of furs produced each year. Both the hare and the lynx showed a regular ten-year cycle, with the peaks in lynx abundance occurring about a year behind the hare's peak abundances. Since the hare is a major food source for the lynx in northern Canada, it is logical to assume that this is a coupled oscillation of population sizes, precisely as predicted by classical predator-prey theory.

Some regular cycles of abundance appear to have evolved as a means of avoiding predation rather than being a direct reduction caused by predation. There is a periodicity in the populations of cicadas and locusts. The hypothesized explanation is that predators cannot reproduce rapidly enough to increase their population sizes quickly in response to the sudden availability of a large food supply. When millions of adult cicadas appear above ground for a few weeks after surviving for seventeen years as nymphs in the soil, predators cannot possibly consume them all: No predator could specialize on adult cicadas unless it also had a seventeen-year cycle.

Several northern bird populations (such as crossbills, grosbeaks, and waxwings) fluctuate dramatically, in some years rising to several times their usual levels. This fluctuation may be a response to changing habitat quality. These bird populations always produce as many eggs as food availability and their natural fecundity allow, even though many offspring will not survive. In a good year, a higher proportion of the offspring survives, and the population experiences an irruption, often leading to intense competition and consequent expansion of the range of the population. In subsequent years, population size returns to preirruption levels. Thus, these fluctuations are entirely consistent with normal density-dependent processes responding to a fluctuating environment.

Irregular Fluctuations
Population fluctuations that occur irregularly or noncyclically often appear to be responses to natural disturbances rather than to density-dependent processes or predator-prey relationships. For example, blue grouse persists at a relatively low level of abundance in coniferous forests until a fire occurs. The species rapidly increases in number following a fire and gradually diminishes again as the forest regenerates over the next several decades.

The population fluctuations of some species are not easily attributed to disturbance or to any other single cause. For example, swarming locusts typically remain at low abundance in a restricted area for several years; then, apparently without warning, they may increase more than a hundredfold and swarm over large areas, consuming large amounts of vegetation. The locust outbreak lasts for several years, then the population declines as rapidly as it initially increased. In the early part of the twentieth century, it was discovered that locusts exhibit two phases: a solitary phase, corresponding to low abundance, and a gregarious phase, corresponding to an outbreak. While it is still not known how locusts transform from one phase to another, it is clear that several stages are involved and that weather conditions seem to initiate a transformation. Moisture seems to be the most important determinant, because of its influence on nymph development and survival, on egg development, and on predator abundance, but wind and plant nitrogen levels have also been implicated. Furthermore, it appears that environmental conditions are only effective in inducing a phase transformation if a certain concentration of locusts already exists and if the existing locusts are adequately sensitive to crowding.

Measuring Fluctuations

There are two parts to the study of population fluctuation: detecting and measuring the pattern of the fluctuation and identifying the underlying causes of the fluctuation. In general, any method designed for measuring population size can be used repeatedly over time to detect fluctuations in the population. Reference to a specialized textbook on ecological sampling techniques is strongly recommended when using any of these methods, in order to assure validity of the sampling for subsequent statistical analysis. The mark-recapture method is commonly used with animal populations. There are many variants of this technique, but they all involve capturing and marking some number of individuals, then releasing them; after some time period appropriate to the study, a second sample is captured and the proportion of marked individuals in the second sample (those that are "recaptured") is recorded. This proportion is used to estimate the size of the population at the time when the individuals were originally marked.

The quadrat method is used primarily with plants and other sessile organisms. Plots (called quadrats) are laid out, either randomly or in some pattern; all individuals within the plots constitute a sample. Quadrats are usually square, but any regular shape may be used. The appropriate size of each quadrat depends on the sizes of organisms to be sampled and on their spatial distribution. If nondestructive sampling techniques are used, the same quadrats may be sampled repeatedly; otherwise, new quadrats must be established for each sampling episode.

A variety of plotless techniques are available for sessile organisms, in lieu of the quadrat method. These techniques were developed to eliminate some of the uncertainties associated with selecting proper quadrat size and location. Most plotless methods locate points on the ground, then measure distances to nearby organisms; each plotless technique identifies the individuals to be measured in a slightly different way.

None of these techniques is adequate by itself to identify the origin or cause of any fluctuation in population size. Experimental manipulation of a population is necessary to elucidate the underlying mechanisms and determining factors. Populations of small, rapidly reproducing species (such as species of *Paramecium* or *Daphnia*) can be manipulated in the laboratory, and hypothesized causes of fluctuation can be tested under controlled conditions. This has been done primarily to develop theoretical predictions regarding environmental conditions (such as temperature, moisture, and humidity), resource limitations and fluctuations, and the effects of predators and competitors.

Identifying Causes of Fluctuation
The most interesting examples of population fluctuation, however, occur over spatial and temporal scales too large to handle in the laboratory. Their underlying mechanisms must be elucidated in the field. Because suites of factors typically produce the complex patterns of population fluctuation observed in nature, an effective field study must include all relevant factors. Generally, the most effective studies have been those that have sought to understand the complete life history of a species. Superb examples include the long-term studies of the wolves of Isle Royal National Park, by David Mech and colleagues, and the equally ambitious studies of the grizzly bears of Yellowstone National Park and surrounding areas by Frank and John Craighead and their many coworkers.

Most equilibrium models of population dynamics are capable of producing regular oscillations that mimic the patterns observed in nature. If the model parameters are properly manipulated, many of these models can produce apparently random fluctuation. More sophisticated models have been constructed that incorporate a mathematical equivalent of random environmental fluctuation, although they usually still assume that a population has a tendency to stabilize and that environmental change simply prevents stabilization. The underlying assumption of almost all these models is that species normally exist at equilibrium. This assumption is consistent with the long-held belief that there is a "balance of nature"— that species exist in harmony with their environment.

If an entire species is considered, perhaps the assumption of equilibrium is warranted, at least for extended periods; yet at the level of the population, fluctuation is the rule—indeed, it may be that extreme fluctuation is the rule. As noted earlier, many populations fluctuate so markedly that they often disappear; they are reestablished only by colonization from large populations within dispersal range. If populations become too small or too isolated from one another, this colonization cannot occur. Additionally, because small populations are more subject to extinction associated with fluctuation, there is additional risk of species extinction if only small populations remain.

The problem of extinction is severe, since habitat destruction is occurring at an unprecedented rate on a global scale. The fragments of intact habitat that remain because of inaccessibility or preservation efforts contain populations that are smaller and more isolated than in the past. If an isolated population fluctuates markedly, resulting in its extinction from a habitat fragment, its replacement by recolonization is unlikely. Furthermore, the genetic variation maintained by a complex population structure within a species is reduced. As the genetic variation within a species is lost, the ability of a species to respond to environmental change is reduced, and extinction of the species is more likely.

Ultimately, if populations normally fluctuate severely enough that they can be expected to become extinct at frequent intervals, then effective conservation requires the maintenance of pathways for exchange and dispersal of individuals among populations within a species. It also requires the preservation of the largest population size possible to allow for normal fluctuation without extinction.

Alan D. Copsey

See also: Biodiversity; Biogeography; Clines, hybrid zones, and introgression; Demographics; Extinctions and evolutionary explosions; Gene flow; Genetic diversity; Genetic drift; Human population growth; Nonrandom mating, genetic drift, and mutation; Population analysis; Population genetics; Population growth; Reproductive strategies; Speciation; Species loss.

Sources for Further Study

Begon, Michael, Martin Mortimer, and David J. Thompson. *Population Ecology: A Unified Study of Animals and Plants*. 3d ed. Cambridge, Mass.: Blackwell Science, 1996.

Gotelli, Nicholas J. *A Primer of Ecology*. 3d ed. Sunderland, Mass.: Sinauer Associates, 2001.

Krebs, Charles J. *Ecology: The Experimental Analysis of Distribution and Abundance*. 4th ed. New York: HarperCollins College Publishers, 1994.

Krebs, Charles J., and J. H. Myers. "Population Cycles in Small Mammals." *Advances in Ecological Research* 8, 1974): 267-399.

Mech, L. David. *The Wolf: The Ecology and Behavior of an Endangered Species.* Garden City, N.Y.: Doubleday, 1970.

Smith, Robert Leo. *Elements of Ecology.* 4th ed. San Francisco: Benjamin/ Cummings, 2000.

Wohrmann, K., and S. K. Jain, eds. *Population Biology: Ecological and Evolutionary Viewpoints.* New York: Springer-Verlag, 1990.

POPULATION GENETICS

Type of ecology: Population ecology

Population genetics is the analysis of genes and genetic traits in populations to determine how much variability exists, what maintains the variability, how selection (natural or controlled) affects a population, and what the mechanisms of evolution are.

Classical genetics deals with the rules of genetic transmission from parents to offspring, developmental genetics deals with the role of genes in development, and molecular genetics looks at the molecular basis of genetic phenomena. Population genetics uses information from all three fields and helps explain why populations are so variable, why some harmful traits are common, why most animals and plants reproduce sexually, how evolution works, why some animals are altruistic in a cutthroat world, and how new species arise.

Mutations and Natural Selection
Simple observation reveals that animals are highly variable. Some dogs are big, others small; some wiry, others big boned; some long-haired, others curly; some with special talents such as herding or retrieving, others with none; and some with diseases or defects, others normal. All of these are the result of various genes combined with environmental influences. Unless an animal is an identical twin, no one else shares that individual's genotype and no one ever will. Population genetics looks at variability in a population and examines its sources and the forces that maintain it.

Variability can come from genetic mutations. For example, about one child in ten thousand is born with dominant achondroplasia (short-legged dwarfism). Some children with the trait inherit the condition from an affected parent, but most have normal parents. They are therefore the result of a new mutation. Many mutations are deleterious and are eventually eliminated from the population by the lowered survival or fertility rates of those who have the mutation, but while they remain in the population, they add to its variability. Occasionally, a seemingly harmful mutation persists, for example, the gene that causes sickle-cell disease, a severe disease characterized by red blood cells that become sickle shaped in certain laboratory tests. The causative gene is recessive, meaning that two copies are needed to produce the anemia, but the disease is very common in some parts of Africa. The harmful, anemia-causing gene persists in the population because if a person has only one gene with the trait rather than two, that gene confers resistance to malaria, the major cause of debility in that part of the world.

Although the genes in these two examples have large and conspicuous effects, the great majority of mutations and the great bulk of genetic variability in the population are the result of a large number of genes with individually small effects, often detected only through statistics. The variability of quantitative traits such as size is due mainly to the cumulative action of many individual genes, each of which produces its small effect. The average size stays roughly constant from generation to generation because individuals who are too large or too small are at a disadvantage. However, such individuals continuously arise from new mutations.

The driving force in evolution is natural selection, that is, the differential survival and fertility of different genotypes. New mutations occur continuously. Most of these are harmful, although usually only mildly so, but a small minority are beneficial. The rules of Mendelian inheritance ensure that the genes are thoroughly scrambled every generation. Natural selection acts like a sieve, retaining those genes that produce favorable phenotypes in the various combinations and rejecting others. Such a process, acting over eons of time, has produced the variety and specific adaptations that can be found throughout the animal kingdom.

The Hardy-Weinberg Law and Selfish Genes

Although evolutionary progress is the result of natural selection, most selection does not accomplish any systematic change. Most selection is directed at maintaining the status quo—eliminating harmful mutations, keeping up with transitory changes in the environment, and eliminating statistical outliers (extremes of variation). Most of the time, evolutionary change is very slow.

In most populations, mating is essentially random in that mates do not choose each other because of the genes they carry. There are exceptions, of course, but for the most part, random mating can be assumed. This permits a great simplification known as the Hardy-Weinberg rule. This rule says that if the proportion of a certain gene, say A, in the population is p and of another, say a, is q, then the three genotypes AA, Aa, and aa are in the proportions p^2, $2pq$, and q^2, respectively. (Remember that p and q are fractions between zero and one.) This is a simple application of elementary probability and the binomial theorem. Furthermore, after a few generations of random mating, genotypes at different loci also equilibrate, which means that the frequency of a composite genotype is the product of the frequencies at the constituent loci. The reason that this is so useful is that the number of genotypes is enormous, but a population can be characterized by a much smaller number of gene frequencies.

Genotypes are transient, but genes may persist unchanged for many generations. This has led the great theorists of population genetics, J. B. S. Haldane and R. A. Fisher in England and Sewall Wright in the United States, to make the primary units the frequency of individual genes and develop theories around this concept, making free use of the simple consequences of random mating. Such a gene-centered view has been described by scholar Richard Dawkins as the "selfish gene." A population can be thought of as a collection of genes, each of which is maximizing its chance of being passed on to future generations. This causes the population to become better adapted because those genes that improve adaptation have the best chance of being perpetuated.

Kin Selection and Selfish Genes

An extension of this notion is kin selection. The concept holds that, to the extent that behavior is determined by genes, individuals should be protective of close relatives because relatives share genes. The fact that brothers and sisters share half their genes should lead a brother to be half as concerned with his sister's survival and reproduction as with his own. Many evolutionists believe that altruistic behavior in various animals, including humans, is the result of kin selection. The degree of self-sacrifice to protect a close relative is proportional to the fraction of shared genes. Parents regularly make sacrifices for their children, and this is what evolutionary theory would predict.

One way in which populations depart from random mating is inbreeding, the mating of individuals more closely related than if they were randomly chosen. Related individuals share one or more ancestors, hence an inbred individual may get two copies of an ancestral gene, one through each parent. In this way, inbreeding increases the proportion of homozygotes. Because many deleterious recessive genes are hidden in the population, inbreeding can have a harmful effect by making genes homozygous. Similarly, if the population is subdivided into local units, mating mostly within themselves, these local units will be more homozygous than if the entire population mated at random. Small subpopulations will be more subject to purely random fluctuations in gene frequencies known as random genetic drifts. Therefore, subdivisions of a population often differ significantly, particularly with respect to unimportant genes.

Views of Evolution: Gene-Centered vs. Genetic Drift

The gene-centered view of evolution is not always accepted. Some evolutionists believe that it is simplistic to view an individual as a bag of genes, each trying to perpetuate itself. They emphasize that genes often interact in complicated ways, and that a theory that deals with only average gene effects is incomplete. Modern theories of evolution take such complications into account.

This different viewpoint has led to a major controversy in evolution, one that has not yet been settled. Wright emphasized that many well-adapted phenotypes depend on genes that interact in very specific ways; two or more genes may be individually harmful but when combined produce a beneficial effect. He argued that selecting genes on the basis of average effects cannot produce such combined effects. He believed that a population subdivided into many partially isolated units provides an opportunity for such interactions. An individual subpopulation, by random drift, might chance upon such a happy gene combination, in which case the whole population can be upgraded by migrants from this subpopulation. Whether evolutionary advance results from gene interactions in subpopulations, from mass selection in largely unstructured populations, or from a combination of both is a question that remains unresolved.

Population genetics theory, along with the techniques of molecular genetics, has greatly deepened our knowledge of historical evolution. Most students of evolution are familiar with tree diagrams of common ancestry that show, for example, birds and mammals branching off from early ancestors. In the past these had to be constructed using external phenotypes and fossils. These techniques for measuring the relatedness of different species and determining their ancestral relations have been replaced by DNA sequencing, which produces much surer results. It has long been suspected that genes can persist for very long evolutionary periods, modifying slightly to perform new, often related but sometimes quite different functions. This belief has been confirmed repeatedly by molecular analysis. The similarity of the DNA sequences between some plant and animal genes is so great as to leave no doubt that they were both derived from a common ancestral gene a billion or more years ago.

Neutral Mutation and Sexual Reproduction

Most gene mutations have very small effects, and the smaller the effect, the less likely it is to be noticed. Molecular techniques have enabled scientists to detect changes in DNA without regard to the traits they cause or whether they have any effect at all. The Japanese geneticist Motoo Kimura has advanced the idea that most evolution at the DNA level is not the result of natural selection but simply the result of mutation and chance, a concept termed neutral mutation. In vertebrates, especially in mammals, most of the DNA has no known function. The functional genes make up a very small fraction of the total DNA. Many scientists believe that most DNA evolution outside the genes—and some within—is the result of changes that are so nearly neutral as to be determined by chance. How large a role random drift plays in the evolution of changes in functional proteins is still not certain.

A few animals and a large number of plants reproduce asexually. Instead of reproducing by using eggs and sperm, the progeny are carbon copies of the parent. Asexual reproduction has obvious advantages. If females could reproduce without males, producing only female offspring like themselves, reproduction would be twice as efficient. However, despite its inherent inefficiency, sexual reproduction is the rule, undoubtedly because of the gene-scrambling process that sex produces. The ability of a species to produce and try out countless gene combinations confers an evolutionary advantage that outweighs the cost of males. Another advantage of gene scrambling is that it permits harmful mutations to be eliminated from the population in groups rather than individually.

Population genetics is also concerned with the processes by which new species arise. Scientists believe that a population somehow becomes divided into two or more isolated groups, separated perhaps by a river, mountain range, or other geographical barrier. Each group then follows its own separate evolutionary course, and the groups' dissimilar environments accentuate their differences. Eventually the two groups evolve so many differences that they are no longer compatible. The products of interspecies crosses, or hybrids, often do not develop normally or are sterile (like the mule). Sometimes the two species do not mate because they are so different.

Research Methods

Population genetics involves theory, observation, and experiment. Population genetics examines how genes are influenced by mutation, selection, population size, migration, and chance. Scientists develop mathematical models that embody these theories and compare the results obtained using the models with data from laboratory experiments or field observations. These genetic models have become more and more sophisticated to take into account complex gene interactions and increasingly realistic population structures. The models are further complicated by efforts to account for random processes. Often the mathematical geneticist relies on computers to perform complex analyses and computations.

One of the simpler models, which makes the assumption that mating is random, is the Hardy-Weinberg principle. If the proportion of gene A in the population is p and that of gene a is q, then the three genotypes AA, Aa, and aa are in the proportions p^2, $2pq$, and q^2, respectively. The proportion of Aa is $2pq$ rather than simply pq because this genotype represents two combinations, maternal A with paternal a and paternal A with maternal a. This principle can be used to predict the frequency of persons with malaria resistance from the incidence of sickle-cell anemia. If one-tenth of the genes are sickle-cell genes and the other nine-tenths are normal, the frequency of two genes coming together to produce an anemic child is 0.1×0.1, or 0.01. The frequency of those resistant to malaria, who have one normal and one sickle-cell gene, is $2 \times 0.1 \times 0.9$, or 0.18. A slight extension of the calculation (using the rates of malaria infection and death from the disease) can be used to estimate the death rate from malaria. Another mathematical model can be formed based on the molecular genetics theory of neutral mutation. A neutral mutation, because it is not influenced by natural selection, has an expected rate of evolution that is equal to the mutation rate. Mathematical models embodying this theory are used to quantitatively predict what will happen in an experiment or what an observational study will find and act as a test of the theory. Neutral mutation theory is quite complicated and requires advanced mathematics.

Observational population genetics consists of studying animals and plants in nature. Evolution rates are inferred from the fossil record. Field observations can determine the frequency of genes in different geographical areas or environments. The frequency of self- and cross-pollination can often be observed directly. Increasingly, DNA analysis, which can detect relationships or alterations that are not visible, is being used to support field observations. For example, molecular markers have been used to determine parentage and relationship. DNA analysis revealed that certain birds that do not reproduce but care for the progeny of others are in fact close relatives, consistent with kin selection theory.

Increasingly, population genetics has begun to rely on experimentation. Plants and animals can be used to study the process of selection, but to save time and reduce costs, most laboratory experiments involve small, rapidly reproducing organisms such as the fruit fly, *Drosophila*. Some of the most sensitive selection experiments have involved the use of a chemostat, a container in which a steady inflow of nutrients and steady outflow of wastes and excess population permits a population to maintain a stable number of rapidly growing organisms, usually bacteria. These permit very sensitive measurements of the effects of mutation. Evolutionary studies that would require eons if studied in large animals or even mice can be completed in a very short time.

Uses of Population Genetics

The greatest intellectual value of population genetics has been to provide a theory of evolution that is explanatory, quantitative, and predictive. Population genetics places knowledge of mutation, gene action, selection, inbreeding, and population structure in a unified framework. It brings together Charles Darwin's theory of evolution by natural selection, Gregor Mendel's laws of inheritance, and molecular genetics to create a coherent picture of how evolution took place and is still occurring.

Population genetics has provided explanations for variability in a population, the prevalence of sexual rather than asexual reproduction, the origin of new species, and behavioral traits such as altruism. It has also provided an understanding of why some harmful diseases are found in the population. Population genetics has been used in animal and plant breeding to create rational selection programs. Using quantitative models, the results of various selection schemes can be compared and the best one chosen.

A particularly telling example of a situation in which population genetics predicted an outcome that has become painfully obvious is the development of resistance to insecticides, herbicides, and antibiotics. As people used these products more and more, the insects, weeds, and bacteria they were trying to eliminate developed resistance, and new products had to be developed to replace those rendered ineffective. The development of resistance represents evolution by natural selection that took place not over hundreds or thousands of years but in just a few years. Probably the most problematic area of resistance is antibiotics because some treatable diseases are again threatening to become beyond the ability of medicine to cure. A major challenge to ecologists, microbiologists, physicians, and population geneticists is how to deal with the increasingly difficult problem of disease-producing microorganisms that are resistant to antibiotics.

James F. Crow

See also: Biodiversity; Biogeography; Clines, hybrid zones, and introgression; Demographics; Extinctions and evolutionary explosions; Gene flow; Genetic diversity; Genetic drift; Human population growth; Nonrandom mating, genetic drift, and mutation; Population analysis; Population fluctuations; Population growth; Reproductive strategies; Speciation; Species loss.

Sources for Further Study

Crow, James F. *Basic Concepts in Population, Quantitative, and Evolutionary Genetics.* New York: W. H. Freeman, 1986.

Dawkins, Richard. *The Blind Watchmaker.* New York: W. W. Norton, 1986.

_____. *The Selfish Gene.* Rev. ed. Oxford, England: Oxford University Press, 1999.

Falconer, Douglas S. *Introduction to Quantitative Genetics.* 4th ed. New York: John Wiley & Sons, 1996.

Fisher, R. A. *The Genetical Theory of Natural Selection.* Rev. ed. New York: Dover, 1958.

Haldane, J. B. S. *The Causes of Evolution.* Reprint. Ithaca, N.Y.: Cornell University Press, 1993.

Hartl, Daniel. *A Primer of Population Genetics.* 3d ed. Sunderland, Mass.: Sinauer Associates, 2000.

Hartl, Daniel, and Andrew Clark. *Principles of Population Genetics.* 3d ed. Sunderland, Mass.: Sinauer Associates, 1997.

Kimura, Motoo. *The Neutral Theory of Molecular Evolution.* Cambridge, England: Cambridge University Press, 1985.

Maynard Smith, John. *The Theory of Evolution.* Cambridge, England: Cambridge University Press, 1993.

Wright, Sewall. *Evolution and the Genetics of Populations.* 4 vols. Chicago: University of Chicago Press, 1968-1978.

POPULATION GROWTH

Type of ecology: Population ecology

Populations typically grow when they are found on sites with abundant resources, and biologists have developed two models to describe growth. In exponential growth, the population is exposed to ideal conditions, and new individuals are added at an ever-increasing rate. Logistic growth recognizes that resources are eventually depleted, however, and that the population density ultimately stabilizes at some level, which is defined as the carrying capacity.

In nature, organisms of a particular species rarely occur by themselves. Instead, they usually exist with other individuals of the same species. Biologists use the term "population" to refer to an aggregation of organisms of a given species that live in the same general location at the same time. In some cases, populations can be well defined, such as herds of cattle or flocks of geese. In other cases, the population is not well defined, often because several species may be found in the same location. For example, a meadow may contain intermingled populations of several species, including daisies, timothy grass, earthworms, and grasshoppers.

Biologists have studied populations for many years. Many of those studies have been conducted to answer three separate but interrelated questions. First, how many individuals are there in a given population at a particular time? Second, how do those numbers change from one time to another? Third, what environmental factors are responsible for any population increases or decreases? Studies have shown that for most species of plants, animals, and microbes, the number of individuals in the population changes over time. Some populations increase steadily, other populations decrease, and still others fluctuate. Thus, populations, when viewed over time, are generally dynamic rather than static.

Population Behaviors

Most populations change so much through time because there is constantly turnover among individuals. That is, new individuals are constantly being born, hatched, or germinated, while others die. Moreover, animals are also able to enter a population by immigration and leave by emigration. Since the number of births and new immigrants hardly ever exactly matches the number of deaths and emigrants, the dynamic nature of populations should not be a surprise. Because changes in population size are common in nature, biologists have tried to understand the changes that are observed. One approach has been to model populations; the model is a simplified graphical or mathematical summary of the actual changes that are occurring in the species of interest. The relationship between a model and the actual population that it represents is similar to that between a map and the area of land that it represents. Because modeling is such an important aspect of population biology, biologists who study population must often have a good background in mathematics.

Perhaps the simplest mode of population behavior is the difference equation, which states that the number of individuals in a population at some specified time in the future is equal to the number at present, plus the number of births, minus the number of deaths, plus the number of new immigrants, minus the number of emigrants. Thus, by knowing how many individuals are on a site at a given time and knowing the usual number of births, deaths, immigrants, and emigrants, one can predict the number of individuals on the site at some future time.

Obviously, the number of births, deaths, immigrants, and emigrants will vary from one place to another and from one time to another. For example, on a site with abundant food and space and with favorable physical conditions for growth and development, births and immigration will be much greater than deaths and emigration. Thus, the population will increase. Conversely, if food or space is limiting or if the physical conditions are more severe, losses to the population through death and emigration will equal or exceed gains through birth and immigration. Thus, the population will remain constant or decline.

Biologists often are concerned about what happens in extreme conditions, because such conditions define the limits within which the population normally operates. When conditions are very bad, a population normally declines rapidly, often to the point of local extinction; when conditions are very good, a population will increase. That increase is attributable to the fact that each individual normally has the capacity to produce many offspring during its lifetime. For example, a woman could produce more than forty children if she conceived every time that she was fertile. Other individual organisms, particularly many invertebrates and plants, can produce hundreds of thousands of offspring in their lifetime.

Birth and Death Rates

At least three different traits influence the reproductive output of a given species. The first is the number of offspring per reproductive period (elephants produce only one child at a time, whereas flies can lay thousands of eggs). The second is the age at first reproduction (most dogs can reproduce when less than three years old, whereas humans do not usually become fertile until they reach an age of thirteen or fourteen). The third is the number of times that an individual reproduces in its lifetime (salmon spawn, that is, lay eggs, only once before they die, whereas chickens lay eggs repeatedly). Even under ideal conditions, death must also be considered when examining population growth. Nearly all organisms have a maximum life span that is determined by their innate physiology and cannot be exceeded, even if they are supplied with abundant food and kept free from disease.

Population biologists frequently express birth and death in the form of rates. This can be done by counting the number of new births and deaths in a population during a predetermined period of time and then dividing by the number of individuals in the population. That will give the per capita (per individual) birth and death rates. For example, suppose that during the course of a year there were thirty births and fifteen deaths in a population of one thousand individuals. That per capita birthrate would be 0.030 and the per capita death rate would be 0.015.

Next, one can subtract the death rate from the birthrate to find the per capita rate of population growth. That rate should be greatest under ideal conditions, when the birthrate is greatest and the death rate is least. That per capita rate of population growth is called the "maximal intrinsic rate of increase" or the "biotic potential" by population biologists, and it is a very important attribute. It is often symbolized as r_{max} or referred to as "little r." Normally, r_{max} is considered an inherent feature of a species. As one might expect, it varies greatly among different types of organisms. For example, r_{max}, expressed per year, is 0.02-1.5 for birds and large mammals, 4-50 for insects and small invertebrates, and as high as 20,000 for bacteria.

Exponential Growth

By knowing the intrinsic rate of increase and the number of organisms in a population, one can predict much about the behavior of a population under ideal conditions. The rate at which the population grows is merely the intrinsic rate of increase (r_{max}) multiplied by the number of individuals in the population. For example, suppose that there are ten individuals in a population whose annual r_{max} is 2. That population would increase by an annual rate of twenty (which would be a healthy increase). Next, suppose that one returned to that population at some later time when the population was fifty individuals. At that point, the annual rate of population increase would be one hundred new individuals (which would be an even healthier increase). If the rate of increase were measured when the population reached five hundred, the annual rate of increase would be one thousand individuals.

Under such circumstances, the population would keep on growing at an ever-increasing rate. That type of growth is called "exponential growth" by population biologists, and it typifies the behavior of many populations when placed under ideal conditions. If the number of individuals in a population undergoing exponential growth is plotted as a function of time, the curve would resemble the letter J. That is, it would be somewhat flat initially, but it would curve upward, and at some point it would be almost vertical. Exponential population growth has been observed to a limited extent in many different kinds of organisms, both in the laboratory and under field conditions. Examples include protozoans, small insects, and birds. It should be obvious, however, that no species could behave in this manner for long. If it did, it would overrun the earth (and indeed the universe) in a matter of time. Instead, population growth is slowed by limited resources, accumulated wastes, behavioral stresses, and/or periodic catastrophes caused by the environment.

Logistic Growth
Biologists have created a second model to account for the behavior of populations under finite resources and have called it logistic growth. If the number of individuals in a population undergoing logistic growth is plotted as a function of time, the curve would resemble a flattened S shape. In other words, the curve would initially be flat, but would then curve upward at a progressively faster rate, much like exponential growth. At some point (called the inflection point), however, the curve would begin to turn to the right and flatten out. Ultimately, the curve would become horizontal, indicating a constant population over time.

An important aspect of pure logistic growth is that the population approaches, but does not exceed, a certain level. That level is called the carrying capacity, and is represented by the symbol K in most mathematical treatments of logistic growth. The carrying capacity is the maximum number of individuals that the environment can support, based on the space, food, and other resources available. When the number of individuals is much fewer than the carrying capacity, the population grows rapidly, much as in exponential growth. As the number increases, however, the rate of population growth becomes much less than the exponential rate. When the number approaches the carrying capacity, new population growth virtually ceases. If the population were to increase above the carrying capacity for some reason, there would be a net loss of organisms from the population.

There are few studies that have documented logistic growth in nature. It would be necessary to watch a species in a habitat from the time of its first introduction until its population stabilized. Such studies are necessarily of a very long duration and thus are not normally conducted. Logistic growth has been found in a number of experimental studies, however, particularly on small organisms, including protozoans, fruit flies, and beetles.

An important aspect of logistic growth is that, as the population increases, the birthrate decreases and the mortality rate increases. Such effects may be attributable to reduced space within which the organism can operate, to less food and other resources, to physiological and behavioral stress caused by crowding, and to increased incidence of disease. Those factors are commonly designated as being density-dependent. They are considered much different from the density-independent factors that typically arise from environmental catastrophes such as flooding, drought, fire, or extreme temperatures. For many years, biologists argued about the relative importance of density-dependent versus density-independent factors in controlling population size. It is now recognized that some species are controlled by density-independent factors, whereas others are controlled by density-dependent factors.

Classically, when a species undergoes logistic growth, the population is ultimately supposed to stabilize at the carrying capacity. Most studies that track populations over the course of time, however, find that numbers actually fluctuate. How can such variability be reconciled with the logistic model? On the one hand, the fluctuations may be caused by density-independent factors, and the logistic equation therefore does not apply. On the other hand, the population may be under density-dependent control, and the logistic model can still hold despite the fluctuations. One explanation for the fluctuations could be that the carrying capacity itself changes over time. For example, a sudden increase in the amount of food available would increase the carrying capacity and allow the population to grow. A second explanation relates to the presence of time lags. That is, a population might not respond immediately to a given resource level. For example, two animals in a rapidly expanding population might mate when the number of individuals is less than the carrying capacity. The progeny, however, might be born several weeks or months later, into a situation in which the population has exceeded the carrying capacity. Thus, there would have to be a decline, leading to the fluctuation.

Approaches to Studying Population Growth

Two main approaches can be used to investigate logistic and exponential population growth among organisms. One approach involves following natural populations in the field; the other involves setting up experimental populations. Each approach has its benefits and drawbacks, and, ideally, both should be employed. To study population growth in the field, it is important to study a species from the time that it first arrives on a site until its population stabilizes. Thus, any species already present are automatically eliminated from consideration unless they are brought to local extinction and a new population is then allowed to recolonize. Population growth studies can be profitably done on sites that are very disturbed and are beginning to fill up with organisms. Examples would be an abandoned farm field or strip mine, a newly created volcanic island, or a new body of water. Moreover, studies could also be done on a species that is purposely introduced to a new site.

In either case, one needs to survey the population periodically to assess the number of individuals that it contains. The size of the population can be determined directly or by employing sampling techniques such as mark-recapture methodology. The number of individuals can then be plotted on a graph (on the y-axis) as a function of time (on the x-axis).

To study population growth in experimental conditions, one sets up an artificial habitat according to the needs of the species in question. For example, investigators have examined population growth in protozoans (unicellular animals) by growing populations in test tubes filled with food dissolved in a known volume of water. Others have grown fruit flies in stoppered flasks. Still others have grown beetles in containers filled with oatmeal or other crushed grain. In those cases, it was typically necessary to replenish the food to keep the population going. Whenever the population was placed into an artificial habitat with a nonrenewable food source, it would generally consume all the food and then die out.

More detailed experiments can be performed to test whether density-dependent mortality is occurring. Such experiments would involve setting up a series of containers with different densities of organisms and then following the mortality of those organisms. In theory, mortality rates should be highest in containers that have the greatest densities of organisms and lowest in containers with the sparsest populations. One could also examine the birthrate in those containers, with the expectation that birthrates should be highest in the sparsest containers and lowest in those that have the most organisms. The investigation should be long enough to allow the population to reach equilibrium at the carrying capacity. For short-lived organisms such as protozoans or insects, that could take days, weeks, or months. For longer-lived organisms such as fish or small mammals, one to several years may be required. For long-lived animals, a truly adequate study may take decades.

Another consideration in studying exponential and logistic population growth is that immigration and emigration should be kept to a minimum.

Thus, organisms that are highly active, such as birds, large mammals, and most flying insects would be extremely difficult to study. Finally, one can set up numerous populations and expose each to a slightly different set of conditions. That would enable the researcher to ascertain which environmental factors are most important in determining the carrying capacity. For example, populations of aquatic invertebrates could be monitored under a range of temperature, salinity, pH, and nutrient conditions.

Research Applications
Since exponential growth is unrealistic in practical terms for almost all populations, its scientific usefulness is limited. The concepts derived from logistic growth, however, have important implications to biologists and nonbiologists alike. One important aspect of logistic growth is that the maximum rate of population growth occurs when the population is about half of the environment's carrying capacity. When populations are very sparse, there are simply too few individuals to produce many progeny. When populations are dense, near the carrying capacity, there is not enough room or other resources to allow for rapid population growth. Based on that relationship, those who must harvest organisms can do so at a rate that allows the population to reestablish quickly.

Those who can apply this concept in their everyday work include wildlife managers, ranchers, and fishermen. Indeed, quotas for hunting and fishing are often set in a way that allows for the population to be thinned sufficiently without depleting it too severely. Unfortunately, there are two problems that biologists must confront when they try to use the logistic model to help manage populations. The first is that it is often difficult to establish the carrying capacity for a given species on a particular site. One reason is that the populations of many species are profoundly affected by density-independent factors, as well as by other species, in highly complex and variable ways. Further, for reasons unclear to most biologists, some species have a maximum rate of population growth at levels well above or well below the level (one half of the carrying capacity) that is normally assumed. Thus, the logistic model typically gives only a very rough approximation for the ideal size of a population. However, the logistic model is useful because it emphasizes that all species have natural limits to the sizes of their populations.

Kenneth M. Klemow

See also: Biodiversity; Biogeography; Clines, hybrid zones, and introgression; Demographics; Extinctions and evolutionary explosions; Gene flow; Genetic diversity; Genetic drift; Human population growth; Nonrandom mating, genetic drift, and mutation; Population analysis; Population fluctuations; Population genetics; Reproductive strategies; Speciation; Species loss.

Sources for Further Study

Brewer, Richard. *The Science of Ecology.* 2d ed. Philadelphia: Saunders College Publishing, 1994.

Elseth, Gerald D., and Kandy D. Baumgardner. *Population Biology.* New York: Van Nostrand, 1981.

Gotelli, Nicholas J. *A Primer of Ecology.* 3d ed. Sunderland, Mass.: Sinauer Associates, 2001.

Hutchinson, G. Evelyn. *An Introduction to Population Ecology.* New Haven, Conn.: Yale University Press, 1978.

Kormondy, Edward J. *Concepts of Ecology.* 4th ed. Englewood Cliffs, N.J.: Prentice-Hall, 1996.

Krebs, Charles J. *Ecology: The Experimental Analysis of Distribution and Abundance.* 5th ed. San Francisco: Benjamin/Cummings, 2001.

Raven, Peter, H., and George B. Johnson. *Biology.* 5th ed. St. Louis: Times Mirror/Mosby, 1999.

Wilson, Edward O., and William Bossert. *A Primer of Population Biology.* Sunderland, Mass.: Sinauer Associates, 1977.

PREDATION

Type of ecology: Behavioral ecology

The relationships among predators and their prey in natural communities are varied and complex. These interactions provide clues as to how natural populations regulate one another, as well as to how to preserve and manage exploited populations more successfully.

Predation is an interaction between two organisms in which one of them, the predator, derives nutrition by killing and eating the other, the prey. Obvious examples include lions feeding on zebras and hawks eating rodents, but predation is not limited to interactions among animals. Birds that feed on seeds are legitimate predators since they are killing individual organisms (embryonic plants) to derive energy. There are a number of species of carnivorous plants, such as sundews and pitcher plants, that capture and consume small animals to obtain nitrogen in habitats that otherwise lack sufficient quantities of that nutrient. Most animals that feed on plants (herbivores) do not kill the entire plant and therefore are not really predators. Exceptions to this generalization are some insects that reach infestation levels, such as gypsy moths or locusts, and can kill the plants upon which they feed. The majority of herbivore-plant associations, however, are more properly described as parasite-host interactions in which the host plant may suffer damage but does not die.

There are special cases in which parasitism and predation may be combined. One of these is the interaction between parasitoid wasps and their hosts, usually flies. Adult female parasitoids attack and inject eggs into fly pupae (the resting stage, during which fly larvae metamorphose into adults), and the larvae of the wasp consume the fly. The adult parasitoid is therefore a parasite, while the larval wasp acts as a predator.

Predator-Prey Interactions

Predator-prey interactions can be divided into two considerations: the effects of prey on predators, and the effects of predators on their prey. Predators respond to changes in prey density (the number of prey in the habitat) in two principal ways. The first is called numerical response, which means that predators change their numbers in response to changes in prey density. This may be accomplished by increasing or decreasing reproduction or by immigrating to or emigrating from a habitat. If prey density increases, predators may immigrate from other habitats to take advantage of this increased resource, or those predators already present may produce more offspring. When prey density decreases, the opposite will occur. Some predators, which are known as fugitive species, are specialized at finding habitats with abundant prey, migrating to them, and reproducing rapidly once they are established.

Cape May warblers are good at finding high densities of spruce bud-worms (a serious pest of conifers) and then converting the energy from their prey into offspring. This strategy allows the birds to persist only because the budworms are never completely wiped out; they are better at dispersing to new habitats than are the birds.

The second response of predators to changing prey density is called functional response. The rate at which predators capture and consume prey depends upon the rate at which they encounter prey, which is a function of prey density. If the predator has a choice of several prey species, it may learn to prefer one of them. If that prey is sufficiently abundant, this situation results in a phenomenon known as switching, the concentration by the predator on the preferred prey. It may entail a change in searching behavior on the part of the predator, such that former prey items will no longer be encountered as frequently.

Animals have evolved a number of defense mechanisms that reduce their probability of being eaten by predators. Spines on horned lizards,

A hungry bear nabs a salmon for lunch as the fish hurls itself upstream to spawn. (Corbis)

threatening displays by harmless snakes, camouflage of many cryptic animals, toxic or distasteful chemicals in insects and amphibians, and simply rapid movement—all are adaptations that may have evolved in response to natural selection by predators. A predator that can learn to prefer one prey item over another is smart enough to learn to avoid less desirable prey. That capability is the basis for a phenomenon known as aposematism among potential prey species which are toxic and/or distasteful to their predators. Aposematic organisms advertise their toxicity by bright coloration, making it easy for predators to learn to avoid them, which in turn saves the prey population from frequent taste-testing. Many species of insects are aposematic. Monarch butterflies are bright orange with black stripes, an easy signal to recognize. They owe their toxicity to the milkweed plant, which they eat as caterpillars. The plant contains cardiac glycosides, which are very toxic. The monarch caterpillar is immune to the poison and stores it in its body so that the adult has a high concentration of it in its wings. If a bird grabs the butterfly in flight, it is likely to get a piece of wing first, and this will teach it not to try orange butterflies in the future.

Some potential prey species that are not themselves toxic have evolved to resemble those that are; these are called Batesian mimics. Viceroy butterflies, which are not toxic, mimic monarchs very closely, so that birds cannot tell them apart. One limit to Batesian mimicry is that mimics can never get very numerous, or their predators will not get a strong enough message to leave them alone. Another kind of mimicry involves mimics that are as toxic as their models. The advantage with this type, Müllerian mimicry, is that the predator has to learn only one coloration signal, which reduces risk for both prey populations. In this relationship, the mimic population does not have to remain at low levels relative to the model population. A third type of mimicry is more insidious—aggressive mimicry, in which a predator resembles a prey or the resource of that prey in order to lure it close enough to capture. There are tropical praying mantises that closely resemble orchid flowers, thus attracting the bees upon which they prey. There are some species of fireflies that eat other species of fireflies, using the flashing light signal of their prey to lure them within range.

The Choice of Prey

What determines predator preference for prey? Since prey are a source of energy for the predator, it might be expected that predators would simply attack the largest prey they could handle. To an extent, this choice holds true for many predators, but there is a cost to be considered. The cost involves the energy a predator must expend to search for, capture, handle, and consume prey. In order to be profitable, a prey item must yield much more energy than it costs. Natural selection should favor reduction in energetic cost relative to energetic gain, the basis for optimal foraging theory. According to this theory, many predators have evolved hunting strategies to optimize the time and energy spent in searching for and capturing prey. Some predators, such as web-building spiders and boa constrictors, ambush their prey. The low energetic cost of sit-and-wait is an advantage in environments that provide plentiful prey. If encounters with prey become less predictably reliable, however, an ambush predator may experience starvation. Spiders can lower their metabolic energy requirements when prey is unavailable, whereas more mobile predators, such as boa constrictors, can simply shift to active searching. Probably because of the likelihood of facing starvation for extended periods of time, ambush predation is more common among animals that do not expend metabolic energy to regulate their body temperatures (ectotherms) than among those that do (endotherms). Some predators, such as wolves and lions, hunt in groups. This allows them to tackle larger (more profitable) prey than if they hunted alone. Solitary hunters generally have to hunt smaller prey.

Natural communities consist of food webs, constructed of links (feeding relationships) among trophic levels. Each prey species is linked to one or more predators. Most predators in nature are generalists with respect to their prey. Spiders, snakes, hawks, lions, and wolves all feed on a variety of prey. Some of these prey are herbivores, but some are themselves predators. Praying mantises eat grasshoppers (herbivores), but they also eat spiders (carnivores) and each other. Thus, generalist predators have a bitrophic niche, in that they occupy two trophic levels at the same time.

Predation and Population Fluctuation

It is an open question whether predators and prey commonly regulate each other's numbers in nature. There are many examples of cyclic changes in abundance over time, in which an increase in prey density is followed by an increase in the numbers of predators, and then the availability of prey decreases, also followed by a decrease in predators. Are predators causing their prey to fluctuate, or are prey responding to some other environmental factor, such as food supply? In the second case, prey may be regulated by food, and in turn, may be regulating predators, but not the reverse.

Predators can sometimes determine the number of prey species that can coexist in a habitat. If a predator feeds on a prey species that could outcompete (competitively exclude) other prey species in a habitat, it may free more resources for those other species. This relationship is known as the keystone effect. Empirical studies have indicated that the number of

Although rare, plant predators exist. This "carnivorous" pitcher plant is formed with a hollow, tubular structure that contains fluid to trap small animals and insects to obtain nitrogen in habitats that otherwise lack sufficient quantities of that nutrient. (Digital Stock)

prey species in some communities is directly related to the intensity of predation (numerical and functional responses of predators) such that at low intensity, few species coexist because of competitive exclusion; at intermediate intensity, the diversity of the prey community is greatest; and at high intensity, diversity decreases because overgrazing begins to eliminate species. This intermediate predation hypothesis depends upon competition among prey species, which is not always the case.

Studying Predation
The central question in the study of predation is: To what extend do predators and their prey regulate one another? Most studies suggest that predators are usually food limited, but the extent to which they regulate their prey is uncertain. It is one thing to observe predators in nature and another to assess their importance to the dynamics of natural communities. Like other aspects of ecology, studies of predation can be descriptive, experimental, and/or mathematical.

At the descriptive level, characteristics of both predator and prey populations are assessed: rates of birth and mortality, age structure, environmental requirements, and behavioral traits. Qualitative and quantitative information of this type is necessary before predictions can be made about the interactions between predator and prey populations. General lack of such information in natural ecosystems is largely responsible for failures at biological control of pests and management of exploited populations.

Experimental studies of predation involve manipulation of predator and/or prey populations. A powerful method of testing the importance of predation is to exclude a predator from portions of its accustomed habitat, leaving other portions intact as experimental controls. In one such experiment, excluding starfish from marine intertidal communities of sessile invertebrates resulted in domination by mussels and exclusion of barnacles and other attached species; in the absence of the predator, one prey species was capable of competitively excluding others. This keystone effect depends upon two factors—that the prey assemblage structure is determined by competition and that the predator preferentially feeds on the species that is the best competitor in the assemblage. Clearly, not all food webs are likely to be structured in this way.

Another method of experimental manipulation is to enhance the numbers of predators in a community. For complex natural communities, both additions and exclusions of predators have revealed direct (depression of prey) and indirect (enhancement) effects. Since generalist predators are bitrophic in nature, they may interact with other carnivores in such a way as to enhance the survival of herbivores that normally would fall victim. In one experiment, adding praying mantises to an insect community resulted in a decrease in spiders and a consequent increase in aphids, normally eaten by these spiders. Such results are not uncommon and contribute to the uncertainty of prediction. Mathematical models have been constructed to depict predator-prey interactions in terms of how each population affects the growth of the other. The simplest of these models, known as the Lotka-Volterra model for the mathematicians who developed it, describes a situation in which prey and predator populations are assumed to be mutually regulating. This model, which was developed for a single prey population and single predator species, has been modified by many workers to provide more realism, but it is far from predicting many competitive situations in complex natural communities.

As with the rest of modern ecology, these different approaches must be blended in order to build a robust picture of how important predators are in natural ecosystems. This knowledge would allow more successful prediction of the outcomes of human intervention and more intelligent management of exploited populations. Predation is a key interaction in natural ecosystems; understanding the nature of this interaction is central to any understanding of nature itself.

Lawrence E. Hurd

541

See also: Balance of nature; Camouflage; Defense mechanisms; Displays; Ethology; Food chains and webs; Hierarchies; Mammalian social systems; Mimicry; Omnivores; Pheromones; Poisonous animals; Territoriality and aggression.

Sources for Further Study

Crawley, M. J., ed. *Natural Enemies: The Population Biology of Predators, Parasites, and Diseases*. Boston: Blackwell Scientific Publications, 1992.

Ehrlich, Paul R., and Anne H. Ehrlich. *Extinction: The Causes and Consequences of the Disappearance of Species*. New York: Random House, 1981.

Fabre, Jean Henri. *The Life of the Spider*. Translated by Alexander Teixeira de Mattos. New York: Dodd, Mead, 1916.

Gause, G. F. *The Struggle for Existence*. Reprint. New York: Dover, 1971.

Hassell, Michael P. *Arthropod Predator-Prey Systems*. Princeton, N.J.: Princeton University Press, 1978.

Krebs, J. R., and N. B. Davies, eds. *Behavioral Ecology: An Evolutionary Approach*. 4th ed. Boston: Blackwell Scientific Publications, 1997.

Levy, Charles Kingston. *Evolutionary Wars: A Three-Billion Years Arms Race*. New York: Basingstoke, 2000.

Mowat, Farley. *Never Cry Wolf*. Boston: Atlantic-Little, Brown, 1963.

PUNCTUATED EQUILIBRIUM VS. GRADUALISM

Types of ecology: Evolutionary ecology; Population ecology; Speciation

According to classical evolutionary theory, new species arise by gradual transfor-mation of ancestral ones. Speciation theory of the 1950's and 1960's, however, pre-dicted that new species arise from small populations isolated from the main popu-lation, where they diverge rapidly.

In 1972, Niles Eldredge and Stephen Jay Gould applied a new concept of speciation to the fossil record, predicting that species should arise sud-denly ("punctuated" by a speciation event) rather than gradually, and then persist virtually unchanged for millions of years in "equilibrium" before becoming extinct or speciating again.

Although Charles Darwin's most influential work was entitled *On the Origin of Species by Means of Natural Selection* (1859), in fact it did not ad-dress the problem in the title. Darwin was concerned with showing that evolution had occurred and that species could change, but he did not deal with the problem of how new species were formed. For nearly a century, no other biologists addressed this problem either. Darwin (and many of his successors) believed that species formed by gradual transformation of ex-isting ancestral species, and this viewpoint (known as gradualism) was deeply entrenched in the biology and paleontology books for a century. In this view, species are not real entities but merely arbitrary segments of con-tinuously evolving lineages that are always in the process of change through time. Paleontologists tried to document examples of this kind of gradual evolution in fossils, but remarkably few examples were found.

The Allopatric Speciation Model

By the 1950's and 1960's, however, systematists (led by Ernst Mayr) began to study species in the wild and therefore saw them in a different light. They noticed that most species do not gradually transform into new ones in the wild but instead have fairly sharp boundaries. These limits are estab-lished by their ability and willingness to interbreed with each other. Those individuals that can interbreed are members of the same species, and those that cannot are of different species. When a population is divided and sep-arated so that formerly interbreeding individuals develop differences that prevent interbreeding, then a new species is formed. Mayr showed that, in nature, large populations of individuals living together (sympatric conditions) interbreed freely, so that evolutionary novelties are swamped out and new species cannot arise.

When a large population becomes split by some sort of barrier so that there are two different populations (allopatric conditions), however, the smaller populations become isolated from interbreeding with the main population. If these allopatric, isolated populations have some sort of unusual gene, their numbers may be small enough that this gene can spread through the whole population in a few generations, giving rise to a new species. Then, when the isolated population is reintroduced to the main population, it has developed a barrier to interbreeding, and a new species becomes established. This concept is known as the allopatric speciation model.

The allopatric speciation model was well known and accepted by most biologists by the 1960's. It predicted that species arise in a few generations from small populations on the fringe of the range of the species, not in the main body of the population. It also predicted that the new species, once it arises on the periphery, will appear suddenly in the main area as a new species in competition with its ancestor. These models of speciation also treated species as real entities, which recognize one another in nature and are stable over long periods of time once they become established. Yet, these ideas did not penetrate the thought of paleontologists for more than a decade after biologists had accepted them.

Eldredge and Gould's Model

In 1972, Niles Eldredge and Stephen Jay Gould proposed that the allopatric speciation model would make very different predictions about species in the fossil record than the prevailing dogma that they must change gradually and continuously through time. In their paper, they described a model of "punctuated equilibrium." Species should arise suddenly in the fossil record (punctuation), followed by long periods of no change (equilibrium, or stasis) until they went extinct or speciated again. They challenged paleontologists to examine their biases about the fossil record and to see if in fact most fossils evolved gradually or rapidly, followed by long periods of stasis.

In the years since that paper, hundreds of studies have been done on many different groups of fossil organisms. Although some of the data were inadequate to test the hypotheses, many good studies have shown quite clearly that punctuated equilibrium describes the evolution of many multicellular organisms. The few exceptions are in the gradual evolution of size (which was specifically exempted by Eldredge and Gould) and in unicellular organisms, which have both sexual and asexual modes of reproduction. Many of the classic studies of gradualism in oysters, heart urchins, horses, and even humans have even been shown to support a model of stasis punctuated by rapid change. The model is still controversial, however, and there are still many who dispute both the model and the data that support it.

Implications

One of the more surprising implications of the model is that long periods of stasis are not predicted by classical evolutionary theory. In neo-Darwinian theory, species are highly flexible, capable of changing in response to environmental changes. Yet, the fossil record clearly shows that most species persist unchanged for millions of years, even when other evidence clearly shows climatic changes taking place. Instead of passively changing in response to the environment, most species stubbornly persist unchanged until they either go extinct, disappear locally, or change rapidly to some new species. They are not infinitely flexible, and no adequate mechanism has yet been proposed to explain the ability of species to maintain themselves in homeostasis in spite of environmental changes and apparent strong natural selection. Naturally, this idea intrigues paleontologists, since it suggests processes that can only be observed in the fossil record and were not predicted from studies of living organisms.

Species Selection

The punctuated equilibrium model has led to even more interesting ideas. If species are real, stable entities that form by speciation events and split into multiple lineages, then multiple species will be formed and compete with one another. Perhaps some species have properties (such as the ability to speciate rapidly, disperse widely, or survive extinction events) that give them advantages over other species. In this case, there might be competition and selection between species, which was called species selection by Steven Stanley in 1975. Some evolutionary biologists are convinced that species selection is a fundamentally different process from that of simple natural selection that operates on individuals. In species selection, the fundamental unit is the species; in natural selection, the fundamental unit is the individual. In species selection, new diversity is created by speciation and pruned by extinction; in natural selection, new diversity is created by mutation and eliminated by death of individuals. There are many other such parallels, but many evolutionary biologists believe that the processes are distinct. Indeed, since species are composed of populations of individuals, species selection operates on a higher level than natural selection.

If species selection is a valid description of processes occurring in nature, then it may be one of the most important elements of evolution. Most evolutionary studies in the past have concentrated on small-scale, or microevolutionary, change, such as the gradual, minute changes in fruit flies or bacteria after generations of breeding. Many evolutionary biologists are convinced, however, that microevolutionary processes are insufficient to explain the large-scale, or macroevolutionary, processes in the evolution of entirely new body plans, such as birds evolving from dinosaurs. In other words, traditional neo-Darwinism says that all evolution is merely microevolution on a larger scale, whereas some evolutionary biologists consider some changes too large for microevolution. They require different kinds of processes for macroevolution to take place. If there is a difference between natural selection (a microevolutionary process) and species selection (a macroevolutionary process), then species selection might be a mechanism for the large-scale changes in the earth's history, such as great adaptive radiations or mass extinctions. Naturally, such radical ideas are still controversial, but they are taken seriously by a growing number of paleontologists and evolutionary biologists. If they are supported by further research, then there may be some radical changes in evolutionary biology.

Reinterpreting the Fossil Record
Determining patterns of evolution requires a very careful, detailed study of the fossil record. To establish whether organisms evolve in a punctuated or gradual mode, many criteria must be met. The taxonomy of the fossils must be well understood, and there must be large enough samples at many successive stratigraphic levels. To estimate the time spanned by the study, there must be some form of dating that allows the numerical age of each sample to be estimated. It is also important to have multiple sequences of these fossils in a number of different areas to rule out the effects of migration of different animals across a given study area. Once the appropriate samples have been selected, then the investigator should measure as many different features as possible. Too many studies in the past have looked at only one feature and therefore established very little. In particular, changes in size alone are not sufficient to establish gradualism, since these phenomena can be explained by many other means. Finally, many studies in the past have failed because they picked one particular lineage or group and selectively ignored all the rest of the fossils in a given area. The question is no longer whether one or more cases of gradualism or punctuation occurs (they both do) but which is predominant among all the organisms in a given study area. Thus, the best studies look at the entire as-

semblage of fossils in a given area over a long stratigraphic interval before they try to answer the question of which tempo and mode of evolution is prevalent.

Since the 1940's, evolutionary biology has been dominated by the neo-Darwinian synthesis of genetics, systematics, and paleontology. In more recent years, many of the accepted neo-Darwinian mechanisms of evolution have been challenged from many sides. Punctuated equilibrium and species selection represent the challenge of the fossil record to neo-Darwinian gradualism and overemphasis on the power of natural selection. If fossils show rapid change and long-term stasis over millions of years, then there is no currently understood evolutionary mechanism for this sort of stability in the face of environmental selection. A more general theory of evolution may be called for, and, in more recent years, paleontologists, molecular biologists, and systematists have all been indicating that such a radical rethinking of evolutionary biology is on the way.

Donald R. Prothero

See also: Adaptations and their mechanisms; Adaptive radiation; Coevolution; Colonization of the land; Convergence and divergence; Dendrochronology; Development and ecological strategies; Evolution: definition and theories; Evolution: history; Evolution of plants and climates; Extinctions and evolutionary explosions; Gene flow; Genetic drift; Genetically modified foods; Isolating mechanisms; Natural selection; Nonrandom mating, genetic drift, and mutation; Paleoecology; Population genetics; Speciation; Species loss.

Sources for Further Study

Bennett, K. D. *Evolution and Ecology: The Pace of Life*. New York: Cambridge University Press, 1997.

Eldredge, Niles. *Time Frames: The Rethinking of Darwinian Evolution and the Theory of Punctuated Equilibria*. New York: Simon & Schuster, 1985.

Eldredge, N., and S. J. Gould. "Punctuated Equilibria: An Alternative to Phyletic Gradualism." In *Models in Paleobiology*, edited by T. J. M. Schopf. San Francisco: Freeman, Cooper, 1972.

Freeman, S., and J. C. Herron. *Evolutionary Analysis*. 2d ed. Upper Saddle River, N.J.: Prentice Hall, 2001.

Gerhart, John. *Cells, Embryos, and Evolution: Toward a Cellular and Developmental Understanding of Phenotypic Variation and Evolutionary Adaptability*. Malden, Mass.: Blackwell Science, 1997.

Gould, Stephen J. "The Meaning of Punctuated Equilibria and Its Role in Validating a Hierarchical Approach to Macroevolution." In *Perspectives on Evolution*, edited by Roger Milkman. Sunderland, Mass.: Sinauer Associates, 1982.

Gould, Stephen J., and Niles Eldredge. "Punctuated Equilibrium: The Tempo and Mode of Evolution Reconsidered." *Paleobiology* 3 (1977): 115-151.

Hoffman, Antoni. *Arguments on Evolution: A Paleontologist's Perspective.* New York: Oxford University Press, 1988.

Levinton, Jeffrey S. *Genetics, Paleontology, and Macroevolution.* 2d ed. New York: Cambridge University Press, 2000.

Mayr, Ernst. *Animal Species and Evolution.* Cambridge, Mass.: Harvard University Press, 1963.

Moller, A. P. *Asymmetry, Developmental Stability, and Evolution.* New York: Oxford University Press, 1997.

RAIN FORESTS

Types of ecology: Biomes; Ecosystem ecology

A forest growing in a region that receives over one hundred inches of rain annually is considered to be a rain forest. Rain forests can be found in both tropical and temperate climates and are noted for their remarkable biodiversity. Thousands of different animal and plant species can be found within only an acre or two of a rain forest.

Rain forests are forests found in regions of the world that receive large amounts of precipitation annually. Rain forests present an incredibly diverse range of habitats, as they exist both at low elevations and high in mountain ranges. Many unusual and seldom-seen creatures inhabit the world's rain forests, including spiders so large they eat small birds, and colorful but highly poisonous tree frogs. The enigmatic sloth, an animal that spends its entire life hanging upside down from tree limbs and moving so slowly that moss grows on its fur, is found in the rain forests of South America.

Although tropical rain forests, such as those in the Amazon River drainage system of South America, are perhaps the best known, rain forests do exist in temperate regions as well. Olympic National Park in the state of Washington preserves a temperate climate rain forest, while much of the coast of British Columbia and southeastern Alaska also receives well over one hundred inches of rain annually. The primary difference between temperate and tropical rain forests is that in a temperate rain forest, often one or two species of trees will become dominant. In the coniferous rain forest of the Pacific Northwest, for example, Douglas fir and western red cedar are the dominant species, while other trees are found in much smaller numbers. In a tropical rain forest, in contrast, several hundred species of trees may grow side by side within a very small geographic area. The majority of trees found in tropical rain forests tend to be broad-leaved, such as the rubber tree, while temperate rain forests are dominated by conifers. The leaves of many plants in rain forests often have a waxy texture or come to a point to help shed water more quickly and prevent the growth of fungi or mold.

Characteristics of Rain Forests

Although rain forests are remarkably diverse, they do share a few characteristics in common. The abundant moisture in a rain forest gives the woodland a lush, fertile appearance. This is particularly true in tropical regions. Even in the understory, close to ground level where light is limited, vegetation may be dense. This appearance of fertility is often deceptive. Dead plant matter decays rapidly in a tropical forest, but the nutrients are used quickly by the numerous competing plants. In addition, the trees in tropical forests are evergreen, which means the litter that does fall to the forest floor does so irregularly, unlike temperate broadleaf forests, where trees lose their leaves annually as the seasons change. Leaves will remain indefinitely on tropical species, such as fig and rubber trees, which is one reason small specimens of these trees are popular as houseplants. As a consequence of this lack of mulch, topsoil is often thin and the root systems of the trees are quite shallow.

One reason tropical rain forests are evergreen is that in the tropics there is little seasonal variation in the hours of daylight. The closer to the equator a forest lies, the less change there is from season to season. In temperate climates, many plants have evolved to bloom, set seeds, or lose their leaves based on the number of hours of sunlight available each day. As the sea-

Although most think of rain forests as tropical ecosystems, many coastal or near-coastal forests in higher latitudes, such as those of the Pacific Northwest, fall into this category as well, with their high rates of precipitation and their various levels of growth. Here, understory and canopy are clearly visible. (PhotoDisc)

sons change annually, plants bloom in the spring or early summer; fruit ripens in the fall, photosynthesis slows, and leaves change color and die. In the tropics, where the number of hours of sun light daily never varies, plants follow a different schedule. Many tropical plants bear new flowers and mature fruit simultaneously. The evergreen foliage and continuous supply of certain fruits has led to the adaptation of some animals to a very restricted diet: koalas, for example, which feed exclusively on eucalyptus leaves, or parakeets that eat only figs. Exceptions to this pattern are the forests where rain fall is seasonal, such as regions of the world like southeast Asia, where much of the rain comes in the form of annual monsoon storms. In those cases, flowering and setting fruit will coincide with the seasonal rains.

Forest Zones

A rain forest can be divided into four zones, each of which has its own distinct characteristics. The lowest level, the forest floor, is often dark and gloomy. Little sunlight penetrates to this level, and there is little air movement. Numerous insects, such as beetles, cockroaches, and termites, live in the decaying litter and provide food for larger animals and birds. Many of the insects, birds, reptiles, and amphibians that live in the lower levels of the rain forest are brightly colored. Scientists speculate that the animals have evolved in this fashion to more easily attract potential mates. Other scientists believe that colors warn potential predators to stay away. In either case, the vivid colors make the animals more easily seen in what is otherwise a dark environment.

Just above the forest floor is the understory. Many of the plants in the understory have large, dark leaves to maximize their light-collecting ability. Because there is little natural air movement within the lower levels of a rain forest due to the canopy blocking any natural breezes, the flowering plants in the understory often have strongly scented or vividly colored flowers to help attract insects or birds to assist with pollination. Lizards, snakes, amphibians such as tree frogs and salamanders, small birds, and mammals as large as the jaguar all call the understory home. The plants found only in the understory seldom exceed fifteen to twenty feet in height. The coffee shrub is an example of a small, shade-tolerant, tropical tree. Until horticulturalists developed strains of coffee for use in plantations where the coffee bushes are the only plants grown, coffee grew naturally in the understory of tropical forests.

The densest layer of plant life is the canopy. High above the rain forest floor, the branches of mature trees form a dense intertwined zone of vegetation extending up as much as 150 feet above the ground. Numerous plants sprout in the crotches of trees, where debris may collect. Tree limbs are festooned with vines and mosses, and bromeliads and orchids grow on the rough bark of tree trunks. Even other trees may start their life cycle a hundred feet above the ground: The strangler figs of Borneo are a relatively shade-intolerant species. A fig seed that lands and sprouts on the ground will probably not survive due to low light levels on the forest floor. Strangler figs have adapted so that their seedlings do best high in the canopy. The figs begin life in the crotches of other trees. The roots of a young fig will gradually creep down the trunk of the tree on which it sprouted. Over time, the strangler fig's roots will completely encircle the host tree and penetrate the forest floor. The fig thrives, but the host tree dies, choked off by the strangler fig. Primates such as gibbons, orangutans, and lemurs spend much of their lives in the canopy, feeding on the fruit of trees such as the strangler fig, as do the sloth and other herbivorous mammals.

The emergent layer of the rain forest consists of the tallest trees, some of which exceed two hundred feet in height. The tops of these trees provide a habitat for large, predatory birds, such as eagles, as well as being home to assorted snakes, monkeys, and other animals. Every layer of the rain forest teems with life, and often what can be found at ground level gives no hint of the diversity that exists two hundred feet above in the tree tops.

Rain Forest Conservation
Many of the trees found in rain forests are valued for their commercial use as lumber, while others have been exploited for their fruits or other products, causing much habitat loss. Tropical hardwoods, such as teak and mahogany, for example, have long been used in construction and in furniture. Teak resists rotting and as a result is often used for products that are going to be exposed to the weather, such as garden furniture. Because teak is desirable as lumber, timber companies are increasingly planting it in plantations for a sustainable yield rather than relying solely on natural forests as a source.

Activists hoping to preserve the tropical rain forest have encouraged indigenous peoples to collect forest products, such as nuts or sap, as a way to create a viable economy while at the same time discouraging industrial clear-cutting of the forest. Native people tap rubber trees in Amazonia, for example, to collect latex. Rubber trees are native to the rain forests of South America, although they are also grown in plantations in other tropical regions of the world, such as Southeast Asia.

The biggest threat to the world's rain forests may not come from commercial logging, however. In many regions of the world, rain forests have fallen victim to population pressures. Forests continue to be clear-cut for agricultural use, even when the farmers and ranchers know the exposed soil's fertility will be quickly exhausted. In some cases, the cleared land becomes an arid wasteland as the tropical sun bakes the soil too hard to absorb rain water. In others, the land is farmed for a year or two and then abandoned. Given enough time, the rain forest may regenerate, but the process will take hundreds of years.

Nancy Farm Männikkö

See also: Biomes: determinants; Biomes: types; Chaparral; Deserts; Forests; Grasslands and prairies; Habitats and biomes; Lakes and limnology; Marine biomes; Mediterranean scrub; Mountain ecosystems; Old-growth forests; Rain forests and the atmosphere; Rangeland; Reefs; Savannas and deciduous tropical forests; Taiga; Tundra and high-altitude biomes; Wetlands.

Sources for Further Study

Bowman, David M. J. S. *Australian Rainforests: Islands of Green in a Land of Fire*. New York: Cambridge University Press, 2000.

Durbin, Kathie. *Pulp Politics and the Fight for the Alaska Rain Forest*. Corvallis: Oregon State University Press, 1999.

Gamlin, Linda, and Anuschka de Rohan. *Mysteries of the Rain Forest*. Pleasantville, N.Y.: Reader's Digest, 1998.

Goldsmith, Frank Barrie. *Tropical Rain Forest: A Wider Perspective*. New York: Chapman & Hall, 1998.

Holloway, M. "Sustaining the Amazon." *Scientific American* 269 (July, 1993): 90-99.

Killman, Wolf, and Lay Thong Hong. "Rubberwood: The Success of an Agricultural By-Product." *Unasylva* 51 (2000): 66-72.

Maser, Chris. *Forest Primeval: The Natural History of an Ancient Forest*. Reprint. Corvallis: Oregon State University Press, 2001.

Tricart, Jean. *The Landforms of the Humid Tropics, Forests, and Savannas*. Translated by Conrad J. Kiewiet de Jonge. New York: St. Martin's Press, 1972.

RAIN FORESTS AND THE ATMOSPHERE

Types of ecology: Biomes; Ecoenergetics; Ecosystem ecology; Global ecology

Because photosynthesis releases large amounts of oxygen into the air, a curtailment of the process by rain-forest deforestation may have negative effects on the global atmosphere.

R ain forests are ecosystems noted for their high biodiversity and high rate of photosynthesis. The rapid deforestation of such areas is of great concern to environmentalists both because it may lead to the extinction of numerous species and because it may reduce the amount of photosynthesis occurring on the earth.

All living things on the earth—plants, animals, and microorganisms—depend on the "sea" of air surrounding them. The atmosphere includes abundant, permanent gases such as nitrogen (78 percent) and oxygen (21 percent) as well as smaller, variable amounts of other gases such as water vapor and carbon dioxide. Organisms absorb and use this air as a source of raw materials and release into it by-products of their life activities.

Cellular Respiration

Cellular respiration is the most universal of the life processes. A series of chemical reactions beginning with glucose and occurring in cytoplasmic organelles called mitochondria, cellular respiration produces a chemical compound called adenosine triphosphate (ATP). This essential substance furnishes the energy cells need to move, to divide, and to synthesize chemical compounds—in essence, to perform all the activities necessary to sustain life. Cellular respiration occurs in plants as well as animals, and it occurs during both the day and the night. In order for the last of the series of chemical reactions in the process to be completed, oxygen from the surrounding air (or water, in the case of aquatic plants) must be absorbed. The carbon dioxide that forms is released into the air.

For cellular respiration to occur, a supply of glucose (a simple carbohydrate compound) is required. Photosynthesis, an elaborate series of chemical reactions occurring in chloroplasts, produces glucose, an organic carbon compound with six carbon atoms. Energy present in light must be trapped by the chlorophyll within the chloroplasts to drive photosynthesis. Therefore, photosynthesis occurs only in plants and related organisms such as algae, and only during the daytime. Carbon dioxide, required as a raw material, is absorbed from the air, while the resulting oxygen is released into the atmosphere. The exchange of gases typically involves tiny openings in leaves, called stomata.

Oxygen Cycle

Oxygen is required for the survival of the majority of microorganisms and all plants and animals. From the surrounding air, organisms obtain the oxygen used in cell respiration. Plants absorb oxygen through the epidermal coverings of their roots and stems and through the stomatal openings of their leaves.

The huge amounts of oxygen removed from the air during respiration must be replaced in order to maintain a constant reservoir of oxygen in the atmosphere. There are two significant sources of oxygen. One involves water molecules of the atmosphere that undergo a process called photodissociation: Oxygen remains after the lighter hydrogen atoms are released from the molecule and escape into outer space.

An aerial view of the rain forest in Guyana. In such tropical forests, the many layers of forest vegetation result in energy from sunlight being efficiently used as it passes downward. The large amounts of oxygen released are available for use not only by the forests but also, because of global air movement, by other ecosystems throughout the world. (PhotoDisc)

The other source is photosynthesis. Chlorophyll-containing organisms release oxygen as they use light as the energy source to split water molecules in a process called photolysis. The hydrogen is transported to the terminal phase of photosynthesis called the Calvin cycle, where it is used as the hydrogen source necessary to produce and release molecules of the carbohydrate glucose. In the meantime, the oxygen from the split water is released into the surrounding air.

Early in the history of the earth, before certain organisms evolved the cellular machinery necessary for photosynthesis, the amount of atmospheric oxygen was very low. As the number and sizes of photosynthetic organisms gradually increased, so did the levels of oxygen in the air. A plateau was reached several million years ago as the rate of oxygen release and absorption reached an equilibrium.

Ozone
Another form of oxygen is ozone. Unlike ordinary atmospheric oxygen, in which each molecule contains two atoms, ozone molecules have three oxygen atoms each. Most ozone is found in the stratosphere at elevations between 10 and 50 kilometers (6 and 31 miles). This layer of ozone helps to protect life on earth from the harmful effects of ultraviolet radiation. Scientists, especially ecologists, are concerned because the amount of ozone has been reduced drastically over the last few decades. Already, an increase in the incidence of skin cancer in humans and a decrease in the efficiency of photosynthesis has been documented. Another concern related to ozone is that of an increase in ozone levels nearer to the ground, where living things are harmed as a result. The formation of ozone from ordinary oxygen within the atmosphere is greatly accelerated by the presence of gaseous pollutants released from industrial processes.

Carbon Cycle
All forms of life are composed of organic (carbon-containing) molecules. Carbohydrates include glucose as well as lipids (fats, oils, steroids, and waxes), proteins, and nucleic acids. The ability of carbon to serve as the backbone of these molecules results from the ability of carbon atoms to form chemical bonds with other carbon atoms and also with oxygen, hydrogen, and nitrogen atoms.

Like oxygen, carbon cycles in a predictable manner between living things and the atmosphere. In photosynthesis, carbon is "fixed" as carbon dioxide in the air (or dissolved in water) is absorbed and converted into carbohydrates. Carbon cycles to animals as they feed on plants and algae. As both green and nongreen organisms respire, some of their carbohydrates are oxidized, releasing carbon dioxide into the air. Each organism must eventually die, after which decay processes return the remainder of the carbon to the atmosphere.

Greenhouse Effect

Levels of atmospheric carbon dioxide have fluctuated gradually during past millennia, as revealed by the analysis of the gas trapped in air bubbles of ice from deep within the earth. In general, levels were lower during glacial periods and higher during warmer ones. After the nineteenth century, levels rose slowly until about 1950 and then much more rapidly afterward. The apparent cause has been the burning of increased amounts of fossil fuels associated with the Industrial Revolution and growing energy demands in its wake. The global warming that is now being experienced is believed by most scientists to be the cause of increased carbon dioxide levels. The greenhouse effect is the term given to the insulating effects of the atmosphere with increased amounts of carbon dioxide. The earth's heat is lost to outer space less rapidly, thus increasing the earth's average temperature.

Forest Ecosystems

The biotic (living) portions of all ecosystems include three ecological or functional categories: producers (plants and algae), consumers (animals), and decomposers (bacteria and fungi). The everyday activities of all organisms involve the constant exchange of oxygen and carbon dioxide between the organisms of all categories and the surrounding atmosphere.

Because they release huge quantities of oxygen during the day, producers deserve special attention. In both fresh and salt water, algae are the principal producers. On land, this role is played by a variety of grasses, other small plants, and trees. Forest ecosystems, dominated by trees but also harboring many other plants, are major systems that produce a disproportionate amount of the oxygen released into the atmosphere by terrestrial ecosystems.

Forests occupy all continents except for Antarctica. A common classification of forests recognizes these principal categories: coniferous (northern evergreen), temperate deciduous, and tropical evergreen, with many subcategories for each. The designation "rain forest" refers to the subcategories of these types that receive an amount of rainfall well above the average. Included are tropical rain forests (the more widespread type) and temperate rain forests. Because of the ample moisture they receive, both types contain lush vegetation that produces and releases oxygen into the atmosphere on a larger scale than do other forests.

Tropical Rain Forests

Tropical rain forests exist at relatively low elevations in a band about the equator. The Amazon basin of South America contains the largest continuous tropical rain forest. Other large expanses are located in western and central Africa and the region from Southeast Asia to Australia. Smaller areas of tropical rain forests occur in Central America and on certain islands of the Caribbean Sea, the Pacific Ocean, and the Indian Ocean. Seasonal changes within tropical rain forests are minimal. Temperatures, with a mean near 25 degrees Celsius, seldom vary more than 4 degrees Celsius. Rainfall each year measures at least 400 centimeters.

Tropical rain forests have the highest biodiversity of any terrestrial ecosystem. Included is a large number of species of flowering plants, insects, and animals. The plants are arranged into layers, or strata. In fact, all forests are stratified but not to the same degree as tropical rain forests. A mature tropical rain forest typically has five layers. Beginning with the uppermost, they are an emergent layer (the tallest trees that project above the next layer); a canopy of tall trees; understory trees; shrubs, tall herbs, and ferns; and low plants on the forest floor.

Several special life-forms are characteristic of the plants of tropical rain forests. Epiphytes are plants such as orchids that are perched high in the branches of trees. Vines called lianas wrap themselves around trees. Most tall trees have trunks that are flared at their bases to form buttresses that help support them in the thin soil.

This brief description of tropical rain forests helps to explain their role in world photosynthesis and the related release of oxygen into the atmosphere. As a result of the many layers of forest vegetation, the energy from sunlight as it passes downward is efficiently utilized. Furthermore, the huge amounts of oxygen released are available for use not only by the forests themselves but also, because of global air movement, by other ecosystems throughout the world. Because of this, tropical rain forests are often referred to as "the earth's lungs."

Temperate Rain Forests

Temperate rain forests are much less extensive than tropical rain forests; they occur primarily along the Pacific Coast in a narrow band from southern Alaska to central California. Growing in this region is a coniferous forest but one with warmer temperatures and a higher rainfall than those to the north and inland. This rainfall of 65 to 400 centimeters per year is much less than that of a tropical rain forest, but is supplemented in the summer by frequent heavy fogs. As a result, evaporation rates are greatly reduced. Because of generally favorable climatic conditions, temperate rain forests, like tropical ones, support a lush vegetation. The rate of photosynthesis and release of oxygen are higher than in most other world ecosystems.

Ecologists and conservationists are greatly concerned about the massive destruction of rain forests. Rain forests are being cut and burned at a rapid rate to plant crops, to graze animals, and to provide timber. The ultimate effect of deforestation of these special ecosystems is yet to be seen.

Thomas E. Hemmerly

See also: Biodiversity; Biomes: determinants; Biomes: types; Communities: ecosystem interactions; Deforestation; Ecosystems: definition and history; Ecosystems: studies; Forests; Geochemical cycles; Global warming; Greenhouse effect; Habitats and biomes; Hydrologic cycle; Nutrient cycles; Ozone depletion and ozone holes; Pollution effects; Rain forests; Savannas and deciduous tropical forests; Slash-and-burn agriculture; Trophic levels and ecological niches.

Sources for Further Study

Laurance, William F., and Richard O. Bierregaard, eds. *Tropical Forest Remnants: Ecology, Management, and Conservation of Fragmented Communities.* Chicago: University of Chicago Press, 1997.

Shipp, Steve. *Rainforest Organizations: A Worldwide Directory of Private and Governmental Entities.* Jefferson, N.C.: McFarland, 1997.

Townsend, Janet G. *Women's Voices from the Rainforest.* New York: Routledge, 1995.

Vandermeer, John. *Breakfast of Biodiversity: The Truth About Rain Forest Destruction.* Oakland, Calif.: Institute for Food and Development Policy, 1995.

RANGELAND

Types of ecology: Agricultural ecology; Biomes; Ecosystem ecology

Open land of a wide variety of types, including grasslands, shrublands, marshes, and meadows as well as some desert and alpine land, is known as rangeland.

Rangeland is a valuable and resilient ecosystem resource that supports considerable plant and animal life. Rangeland generally refers to a kind of land rather than a use of that land. The Society for Range Management defines rangelands as "land on which the native vegetation (climax or natural potential) is predominantly grasses, grass-like plants, forbs, or shrubs." Rangeland "includes lands revegetated naturally or artificially" as well as "natural grasslands, savannas, shrublands, most deserts, tundra, alpine communities, coastal marshes and wet meadows."

Rangelands usually have some limitation for intensive agriculture, such as low and erratic precipitation, lack of soil fertility, shallow or rocky soil, or steep slopes. In addition to livestock grazing, rangelands serve multiple-use functions such as providing recreational opportunities, watersheds, mining locations, and habitat for many animal species. Renewable natural resources associated with rangelands are plants and animals (and, in some senses, water). Nonrenewable resources include minerals and other extractable materials.

Location and Characteristics

Rangelands are extensive and extremely variable. As defined by the Society for Range Management, they occupy more than 50 percent of the world's total land surface and about 1 billion acres in the United States alone. Rangelands are home to nomadic herders on nearly every continent. They vary from high-elevation alpine tundra and high-latitude Arctic tundra to tropical grasslands. The tall-grass prairies in the United States (now mostly plowed for intensive agriculture) and the rich grasslands of eastern Africa are among the most productive.

Rangelands grade into woodlands and forests as woody species and trees become more abundant. Some forests are grazed by wild and domestic animals, and the distinction between rangeland and forest is often not clear. The other difficult distinction is between rangeland and pastureland. Pastureland is generally improved by seeding, fertilization, or irrigation, whereas rangelands support native plants and have little intensive improvement.

In the United States, rangeland improvements during the twenty years following World War II often included brush control, grazing management, and seeding, but rangelands were not irrigated. After the 1970's, when fuel costs increased and environmental concerns about pesticide use increased, brush control practices were reduced considerably. Today environmental concerns include rangeland degradation from overgrazing, especially on riparian vegetation along streams, and concern for endangered animal and plant species. These issues have become controversial in the United States.

Rangelands as Ecosystems

Rangelands constitute natural ecosystems with nonliving environmental factors such as soil and climatic factors. Life-forms are primary producers (grasses, forbs, and shrubs), herbivores (livestock; big game animals such as deer and bison; and many rodents and insects), carnivores (such as coyotes, bears, and eagles), and decomposers (fungi and bacteria) that break down organic matter into elements that can be utilized by plants. Plants convert carbon dioxide and water into complex carbohydrates, fats, and proteins that nourish animals feeding on the plants.

Horses graze amid sagebrush in Pinedale, Wyoming. In addition to livestock grazing, rangelands serve multiple-use functions, providing recreational opportunities, watersheds, and habitat for many animal and plant species. (AP/Wide World Photos)

Individual chemical elements are circulated throughout the various components. Many of these elements are present in the soil, including phosphorus, magnesium, potassium, and sulfur. Nitrogen, on the other hand, is present in large amounts in the atmosphere but must be converted (fixed) into forms that can be utilized by plants before it can be cycled. Energy is fixed through the process of photosynthesis and transformed to forms useful for the plants, then the animals that feed on plants.

When chemicals are taken up by plant roots from the soil, they become available to a wide group of herbivores, from small microbes to large ungulates. Eventually nutrients are passed on to organisms at higher trophic levels (omnivores and carnivores). Both plant and animal litter is eventually broken down by decomposers—bacteria, fungi, and other soil organisms—and returned to the soil or, in the case of carbon or nitrogen, given off to the atmosphere.

However, energy is degraded at each step along the way; energy is transferred but not cycled. Grazing animals on rangelands influence plants by removing living tissue, by trampling, and by altering competitive relations with other plants. Large grazing animals tend to compact the soil, reducing infiltration and increasing surface runoff.

Rangeland Dynamics

Rangelands vary considerably with time. Scientists are gaining a better understanding of some factors related to rangeland change. Pollen records and, in the southwestern United States, packrat middens have been used to reconstruct past climate and vegetational conditions. Some areas have become drier and others more mesic. The formation and retreat of glaciers influenced climatic patterns and soil development. A recent general trend in many rangelands is an increase in woody plants at the expense of grasses. Many factors are probably responsible for these shifts, but fire control, overgrazing, climatic shifts, introduction of exotic species, and influence of native animals are likely causal agents.

Rangelands are being threatened by encroachment from crop agriculture as worldwide development increases. Nomadic herders traditionally met periodic drought conditions by having the flexibility to move to areas not impacted by drought. Now, with area lost to livestock grazing and other political restrictions, herders are often forced to maintain higher livestock numbers to support those directly dependent on livestock. Despite various kinds of disturbances and stresses on rangelands, these areas have supported many large grazing animals and people for centuries.

Rex D. Pieper

See also: Forest management; Grasslands and prairies; Grazing and over-grazing; Multiple-use approach.

Sources for Further Study

Heady, Harold F., and R. Dennis Child. *Rangeland Ecology and Management.* 2d ed. Boulder, Colo.: Westview, 2000.

Holechek, Jerry L., Rex D. Pieper, and Carlton H. Herbel. *Range Management: Principles and Practices.* 4th ed. Upper Saddle River, N.J.: Prentice Hall, 2001.

Jacobs, Lynn. *Waste of the West: Public Lands Ranching.* Tucson, Ariz.: L. Jacobs, 1991.

Longworth, John W., and Gregory J. Williamson. *China's Pastoral Region: Sheep and Wool, Minority Nationalities, Rangeland Degradation, and Sustainable Development.* Wallingford, England: CAB International, 1993.

Owen, Oliver S., Daniel D. Chiras, and John P. Reganold. *Natural Resource Conservation.* 7th ed. Upper Saddle River, N.J.: Prentice Hall, 1998.

Sayre, Nathan F. *The New Ranch Handbook: A Guide to Restoring Western Rangelands.* Santa Fe, N.Mex.: Quivera Coalition, 2001.

REEFS

Types of ecology: Aquatic and marine ecology; Biomes; Ecosystem ecology

Reefs are among the oldest known communities, existing at least 2 billion years ago. They exert considerable control on the surrounding physical environment, influencing turbulence levels and patterns of sedimentation. Ancient reefs are often important hydrocarbon reservoirs.

"True" Reefs vs. Reeflike Structures

Reefs or reeflike structures are among the oldest known communities, extending back more than 2 billion years into the earth's history. These earliest reefs were vastly different in their biotic composition and physical structure from modern reefs, which are among the most diverse of biotic communities and display amazingly high rates of biotic productivity (carbon fixation) and calcium carbonate deposition, despite their existence in a virtual nutrient "desert." Reefs are among the few communities to rival the power of humankind as a shaper of the planet. The Great Barrier Reef of Australia, for example, forms a structure some 2,000 kilometers in length and up to 150 kilometers in width.

It is necessary to distinguish between "true," or structural, reefs and reeflike structures or banks. Reefs are carbonate structures that possess an internal framework. The framework traps sediment and provides resistance to wave action; thus, reefs can exist in very shallow water and may grow to the surface of the oceans. Banks are also biogenically produced but lack an internal framework. Thus, banks are often restricted to low-energy, deep-water settings. "Bioherm" refers to moundlike carbonate buildups, either reefs or banks, and "biostrome" to low, lens-shaped buildups.

Reef Classification

Modern reefs are classified into several geomorphic types: atoll, barrier, fringing, and patch. Many of these may be further subdivided into reef crest or flat, back-reef or lagoon, and fore-reef zones. Atoll reefs are circular structures with a central lagoon, thought to form on subsiding volcanic islands. Barrier reefs are elongate structures that parallel coastlines and possess a significant lagoon between the exposed reef crest and shore. These often occur on the edges of shelves that are uplifted by faulting. Fringing reefs are elongate structures paralleling and extending seaward from the coastline that lack a lagoon between shore and exposed reef crest. Patch reefs are typically small, moundlike structures, occurring isolated on shelves or in lagoons. The majority of fossil reefs would be classified as patch reefs, although many examples of extensive, linear, shelf-edge trends are also known from the geologic record.

Reefs form one of the most distinctive and easily recognized sedimentary facies (or environments). In addition to possessing a characteristic fauna consisting of corals, various algae, and stromatoporoids, they are distinguished by a massive (nonlayered) core that has abrupt contacts with adjacent facies. Associated facies include flat-lying lagoon and steeply inclined fore-reef talus, the latter often consisting of large angular blocks derived from the core. The reef core is typically a thick unit relative to adjacent deposits. The core also consists of relatively pure calcium carbonate with little contained terrigenous material.

Reef Environments

Modern reefs are restricted to certain environments. They occur abundantly only between 23 degrees north and south latitudes and tend to be restricted to the western side of ocean basins, which lack upwelling of cold bottom waters. This restriction is based on temperature, as reefs do not flourish where temperatures frequently reach below 18 degrees Celsius. Reef growth is largely restricted to depths greater than 60 meters, as there is insufficient penetration of sunlight below this depth for symbiont-bearing corals to flourish. Reefs also require clear waters lacking suspended terrigenous materials, as these interfere with the feeding activity of many reef organisms and also reduce the penetration of sunlight. Finally, most reef organisms require salinities that are in the normal oceanic range. It appears that many fossil reefs were similarly limited in their environmental requirements.

Some of the most striking features of modern reefs include their pronounced zonation, great diversity, and high productivity and growth rates. Reefs demonstrate a strong bathymetric (depth-related) zonation. This zonation is largely mediated through depth-related changes in turbulence intensity and in the quantity and spectral characteristics (reds are absorbed first, blues last) of available light. Shallow (1- to 5-meter) fore-reef environments are characterized by strong turbulence and high light intensity and possess low-diversity assemblages of wave-resistant corals, such as the elk-horn coral, *Acropora palmata*, and crustose red algae.

With increasing depth (10-20 meters), turbulence levels decrease and coral species diversity increases, with mound and delicate branching colonies occurring. At greater depths (30-60 meters), corals assume a flattened, platelike form in an attempt to maximize surface area for exposure to ambient light. Sponges and many green algae are also very important over this range. Finally, corals possessing zooxanthellae, which live in the coral tissues and provide food for the coral host, are rare or absent below 60 meters because of insufficient light. Surprisingly, green and red calcareous algae extend to much greater depths (100-200 meters), despite the very low light intensity (much less than 1 percent of surface irradiance). Sponges are also important members of these deep reef communities.

Diveristy of Life-Forms
Coral reefs are among the most diverse of the earth's communities; however, there is no consensus on the mechanism(s) behind the maintenance of this great diversity. At one time, it was believed that reefs existed in a low-disturbance, highly stable environment, which allowed very fine subdivision of food and habitat resources and thus permitted the coexistence of a great number of different species. Upon closer inspection, however, many reef organisms appear to overlap greatly in food and habitat requirements. Also, it has become increasingly apparent that disturbance, in the form of disease, extreme temperatures, and hurricanes, is no stranger to reef communities.

Coral reefs exhibit very high rates of productivity (carbon fixation), which is a result of extremely tight recycling of existing nutrients. This is necessary, as coral reefs exist in virtual nutrient "deserts." Modern corals exhibit high skeletal growth rates, up to 10 centimeters per year for some branching species. Such high rates of skeletal production are intimately related to the symbiosis existing between the hermatypic or reef-building scleractinian corals (also gorgonians and many sponges) and unicellular algae or zooxanthellae. Corals that, for some reason, have lost their zooxanthellae or that are kept in dark rooms exhibit greatly reduced rates of skeleton production.

In addition to high individual growth rates for component taxa, the carbonate mass of the reefs may grow at a rate of some 2 meters per 1,000 years, a rate that is much higher than that of most other sedimentary deposits. This reflects the high productivity or growth rates of the component organisms and the efficient trapping of derived sediment by the reef frame. Although the framework organisms, most notably corals, are perhaps the most striking components of the reef system, the framework represents only 10-20 percent of most fossil reef masses. The remainder of the reef mass consists of sedimentary fill derived from the reef community through a combination of biosynthesis (secretion) and bioerosion (breaking down) of calcium carbonate. An example of the relative contributions of reef organisms to sediment can be found in Jamaica, where shallow-water, back-

This underwater close-up of a portion of Australia's Great Barrier Reef displays the diversity of species supported by this marine ecosystem. The Great Barrier Reef forms a structure some 2,000 kilometers in length and up to 150 kilometers in width. (Corbis)

reef sediment consists of 41 percent coral, 24 percent green calcareous algae, 13 percent red calcareous algae, 6 percent foraminifera, 4 percent mollusks, and 12 percent other grains. The most important bioeroders are boring sponges, bivalves, and various "worms," which excavate living spaces within reef rock or skeletons, and parrot fish and sea urchins, which remove calcium carbonate as they feed upon surface films of algae.

Types of Reef Communities

A diversity of organisms has produced reef and reeflike structures throughout the earth's history. Several distinct reef community types have been noted, as well as four major "collapses" of reef communities. The oldest reefs or reeflike structures existed more than 2 billion years ago during the Precambrian eon. These consisted of low-diversity communities dominated by soft, blue-green algae, which trapped sediment to produce layered, often columnar structures known as stromatolites.

During the Early Cambrian period, blue-green algae were joined by calcareous, conical, spongelike organisms known as archaeocyathids, which persisted until the end of the Middle Cambrian. Following the extinction of the archaeocyathids, reefs again consisted only of blue-green algae until the advent of more modern reef communities in the Middle Ordovician period. These reefs consisted of corals (predominantly tabulate and, to a much lesser extent, rugose corals), red calcareous algae, bryozoans (moss animals), and the spongelike stromatoporoids. This community type persisted through the Devonian period, at which time a global collapse of reef communities occurred.

The succeeding Carboniferous period largely lacked reefs, although algal and crinoidal (sea lily) mounds were common. Reefs again occurred in the Permian period, consisting mainly of red and green calcareous algae, stromatolites, bryozoans, and chambered calcareous sponges known as sphinctozoans, which resembled strings of beads. These reefs were very different from those of the earlier Paleozoic era; in particular, the tabulates and stromatoporoids no longer played an important role. The famous El Capitan reef complex of West Texas formed during this interval. The Paleozoic era ended with a sweeping extinction event that involved not only reef inhabitants but also other marine organisms.

After the Paleozoic extinctions, reefs were largely absent during the early part of the Mesozoic era. The advent of modern reefs consisting of scleractinian corals and red and green algae occurred in the Late Triassic period. Stromatoporoids once again occurred abundantly on reefs during this interval; however, the role of the previously ubiquitous blue-green algal stromatolites in reefs declined. Late Cretaceous reefs were often dominated by conical, rudistid bivalves that developed the ability to form frameworks and may have possessed symbiotic relationships with algae, as do many modern corals. Rudists, however, became extinct during the sweeping extinctions that occurred at the end of the Cretaceous period. The reefs that were reestablished in the Cenozoic era lacked stromatoporoids and rudists and consisted of scleractinian corals and red and green calcareous algae. This reef type has persisted, with fluctuations, until the present.

Study of Modern Reefs

Modern reefs are typically studied while scuba (*self-contained underwater breathing apparatus*) diving, which enables observation and sampling to a depth of approximately 50 meters. Deeper environments have been made accessible through the availability of manned submersibles and unmanned, remotely operated vehicles that carry mechanical samplers and still and video cameras. The biological compositions of reef communities are determined by census (counting) methods commonly employed by plant ecologists. Studies of symbioses, such as that between corals and their zooxanthellae, employ radioactive tracers to determine the transfer of products between symbiont and host. Growth rates are measured by staining the calcareous skeletons of living organisms with a dye, such as Alizarin red, and then later collecting and sectioning the specimen and measuring the amount of skeleton added since the time of staining. Another method for determining growth is to X-ray a thin slice of skeleton and then measure and count the yearly growth bands that are revealed on the X-radiograph. Variations in growth banding reflect, among other factors, fluctuations in ocean temperature.

Reef sediments, which will potentially be transformed into reef limestones, are examined through sieving, X-ray diffraction, and epoxy impregnation and thin-sectioning. Sieving enables the determination of sediment texture, the relationships of grain sizes and abundance (which will reflect environmental energy and the production), and erosion of grains through biotic processes. X-ray diffraction produces a pattern that is determined by the internal crystalline structure of the sediment grains. As each mineral possesses a unique structure, the mineralogical identity of the sediment may be determined. Thin sections of embedded sediment or lithified rock are examined with petrographic microscopes, which reveal the characteristic microstructures of the individual grains. Thus, even highly abraded fragments of coral or algae may be identified and their contributions to the reef sediment determined.

Because of their typically massive nature, fossil reefs are usually studied by thin-sectioning of lithified rock samples collected either from surface exposures or well cores. Reef limestones that have not undergone extensive alteration may be dated through carbon 14 dating, if relatively young, or through uranium-series radiometric dating methods.

Reefs as Ecological Laboratories

Modern reefs serve as natural laboratories, enabling the scientists to witness and study phenomena, such as carbonate sediment production, bioerosion, and early cementation, that have been responsible for forming major carbonate rock bodies in the past. The study of cores extracted from centuries-old coral colonies shows promise for deciphering past climates and perhaps predicting future trends. This is made possible by the fact that the coral skeleton records variations in growth that are related to ocean temperature fluctuations. The highly diverse modern reefs also serve as ecological laboratories for testing models on the control of community structure. For example, the relative importance of stability versus disturbance and recruitment versus predation in determining community structure is being studied within the reef setting.

Modern reefs are economically significant resources, particularly for many developing nations in the tropics. Reefs and the associated lagoonal sea-grass beds serve as important nurseries and habitats for many fish and invertebrates. The standing crop of fish immediately over reefs is much higher than that of adjacent open shelf areas. Reef organisms may one day provide an important source of pharmaceutical compounds, such as prostaglandins, which may be extracted from gorgonians (octocorals). In addition, research has focused upon the antifouling properties exhibited by certain reef encrusters. Reefs also provide recreational opportunities for snorkelers and for scuba divers, a fact that many developing countries are utilizing to promote their tourist industries. Finally, reefs serve to protect shorelines from wave erosion.

Because of the highly restricted environmental tolerances of reef organisms, the occurrence of reefs in ancient strata enables fairly confident estimation of paleolatitude, temperature, depth, salinity, and water clarity. In addition, depth- or turbulence-related variation in growth form (mounds in very shallow water, branches at intermediate depths, and plates at greater depths) enables even more precise estimation of paleobathymetry or turbulence levels. Finally, buried ancient reefs are often important reservoir rocks for hydrocarbons and thus are important economic resources.

W. David Liddell

See also: Defense mechanisms; Marine biomes; Ocean pollution and oil spills; Phytoplankton.

Sources for Further Study

Birkeland, Charles, ed. *Life and Death of Coral Reefs*. New York: Chapman and Hall, 1997.

Cousteau, Jacques-Yves, and Philippe Diolé. *Life and Death in a Coral Sea*. Translated by J. F. Bernard. Garden City, N.Y.: Doubleday, 1971.

Darwin, Charles. *The Structure and Distribution of Coral Reefs*. 1851. Reprint. Berkeley: University of California Press, 1962.

Davidson, Osha Gray. *The Enchanted Braid: Coming to Terms with Nature on the Coral Reef*. New York: Wiley, 1998.

Frost, S. H., M. P. Weiss, and J. B. Sanders, eds. *Reefs and Related Carbonates: Ecology and Sedimentology*. Tulsa, Okla.: American Association of Petroleum Geologists, 1977.

Goreau, Thomas F., et al. "Corals and Coral Reefs." *Scientific American* 241 (August, 1979): 16.

Jones, O. A., and R. Endean, eds. *Biology and Geology of Coral Reefs*. 4 vols. New York: Academic Press, 1973-1977.

Kaplan, Eugene H. *A Field Guide to Coral Reefs of the Caribbean and Florida, Including Bermuda and the Bahamas*. Boston: Houghton Mifflin, 1988.

Keating, Barbara H., et al., eds. *Seamounts, Islands, and Atolls*. Washington, D.C.: American Geophysical Union, 1987.

Köhler, Annemarie, and Danja Köhler. *The Underwater Explorer: Secrets of a Blue Universe*. New York: Lyons Press, 1997.
National Geographic Society. *Jewels of the Caribbean Sea: A National Geographic Special*. Video. Washington, D.C.: Author, 1994.
Newell, Norman D. "The Evolution of Reefs." *Scientific American* 226 (June, 1972): 12, 54-65.
Wood, Rachel. *Reef Evolution*. New York: Oxford University Press, 1999.

REFORESTATION

Type of ecology: Restoration and conservation ecology

Reforestation is the growth of new trees in an area that has been cleared for human activities. It can occur naturally or be initiated by people.

Many areas of the eastern United States, such as the New England region, reforested naturally in the nineteenth and early twentieth centuries after farmland that had been abandoned was allowed to lie fallow for decades. After an area has been logged, environmentalists, as well as the commercial logging industry, advocate planting trees rather than waiting for natural regrowth because the process of natural regeneration can be both slow and unpredictable. In natural regeneration, the mixture of trees in an area may differ significantly from the forest that preceded it. For example, when nineteenth century loggers clear-cut the white pine forests of the Great Lakes region, many logged-over tracts grew back primarily in mixed hardwoods.

Land that has been damaged by industrial pollution or inefficient agricultural practices sometimes loses the ability to reforest naturally. In some regions of Africa, soils exposed by slash-and-burn agriculture contain high levels of iron or aluminum oxide. Without a protective cover of vegetation, even under cultivation, soil may undergo a process known as laterization. Laterite is a residual product of rock decay that makes soil rock-hard. Such abandoned farmland is likely to remain barren of plant life for many years. In polluted areas such as former mining districts, native trees may not be able to tolerate the toxins in the soil; in these cases, more tolerant species must be introduced.

Safeguarding Timber Resources

Reforestation differs from tree farming in that the goal of reforestation is not always to provide woodlands for future harvest. Although tree farming is a type of reforestation (trees are planted to replace those that have been removed), generally only one species of tree is planted, with explicit plans for its future harvest. The trees are seen first as a crop and only incidentally as wildlife habitat or a means of erosion control.

As foresters have become knowledgeable about the complex interactions within forest ecosystems, however, tree farming methods have begun to change. Rather than monocropping (planting only one variety of tree), the commercial forest industry has begun planting mixed stands. Trees that possessed no commercial value, once considered undesirable weed trees, are now recognized as nitrogen fixers necessary for the healthy growth of other species. In addition to providing woodlands for possible use in commercial forestry, goals of reforestation include wildlife habitat restoration and the reversal of environmental degradation.

Early Efforts
Reforestation to replace trees removed for commercial purposes has been practiced in Western Europe since the late Middle Ages. English monarchs, including Queen Elizabeth I, realized that forests were a vanishing resource and established plantations of oaks and other hardwoods to ensure a supply of ship timbers. Similarly, Sweden created a corps of royal foresters to plant trees and watch over existing woodlands. These early efforts at reforestation were inspired by the reduction of a valuable natural resource. By the mid-nineteenth century it was widely understood that the removal of forest cover contributes to soil erosion, water pollution, and the disappearance of many species of wildlife.

Clear-cutting removes all the trees on a tract of land, leaving none standing. At one time a standard practice in logging, it has become one of the most controversial harvesting techniques used in modern logging. With its windrows of slash and debris, a clear-cut tract of land may appear as though a catastrophic event has devastated the landscape. (PhotoDisc)

The laborious process of planting new trees, combined with the time required for those trees to grow, highlights the need to eliminate clear-cutting as a means of harvesting timber. Reforestation also requires careful reconstruction of the forest community in all its variety, in order to provide habitat for many forms of wildlife. Simple "tree farming," by contrast, is aimed at growing stands of a single species of tree for later harvesting. (PhotoDisc)

Ecological and Environmental Aspects

Water falling on hillsides made barren by clear-cutting timber washes away topsoil and causes rivers to choke with sediment, killing aquatic life. Without trees to slow the flow of water, rain can also run off slopes too quickly, causing rivers to flood. For many years, soil conservationists advocated reforestation as a way to counteract the ecological damage caused by erosion.

In the mid-twentieth century, scientists established the vital role that trees, particularly those in tropical rain forests, play in removing carbon dioxide from the earth's atmosphere through the process of photosynthesis. Carbon dioxide is a greenhouse gas: It helps trap heat in the atmosphere. As forests disappear, the risk of global warming—caused in part by an increase in the amount of carbon dioxide in the atmosphere—becomes greater. Since the 1980's, scientists and environmental activists concerned about global warming have joined foresters and soil conservationists in urging that for every tree removed anywhere, whether to clear land for development or to harvest timber, replacement trees be planted. As the area covered by tropical rain forests shrinks in size, the threat of irreversible damage to the global environment becomes greater.

Reforestation Programs

In 1988 American Forests, an industry group, established the Global ReLeaf program to encourage reforestation efforts in an attempt to combat global warming. In addition to supporting reforestation efforts by government agencies, corporations, and environmental organizations, Global ReLeaf and similar programs encourage people to practice reforestation in their own neighborhoods. Trees serve as a natural climate control, helping to moderate extremes in temperature and wind. Trees in a well-landscaped yard can reduce a homeowner's energy costs by providing shade in the summer and serving as a windbreak during the winter. Global ReLeaf is one of many programs that support reforestation efforts.

Arbor Day, an annual day devoted to planting trees for the beautification of towns or the forestation of empty tracts of land, was established in the United States in 1872. The holiday originated in Nebraska, a prairie state that seemed unnaturally barren to homesteaders used to eastern woodlands. Initially emphasizing planting trees where none had existed before, Arbor Day is observed in U.S. public schools to educate young people about the importance of forest preservation. Organizations such as the National Arbor Day Foundation provide saplings (young trees) to schools and other organizations for planting in their own neighborhoods.

Nancy Farm Männikkö

See also: Biodiversity; Biomass related to energy; Conservation biology; Deforestation; Endangered plant species; Erosion and erosion control; Forest fires; Forest management; Forests; Grazing and overgrazing; Integrated pest management; Multiple-use approach; Old-growth forests; Rain forests; Restoration ecology; Species loss; Sustainable development; Urban and suburban wildlife; Wildlife management.

Sources for Further Study

Cherrington, Mark. *Degradation of the Land*. New York: Chelsea House, 1992.

TreePeople, with Andy Lipkis and Katie Lipkis. *The Simple Act of Planting a Tree: A Citizen Forester's Guide to Healing Your Neighborhood, Your City, and Your World*. Los Angeles: St. Martin's Press, 1990.

Weiner, Michael. *Plant a Tree: Choosing, Planting, and Maintaining This Precious Resource*. New York: Wiley, 1992.

REPRODUCTIVE STRATEGIES

Types of ecology: Behavioral ecology; Population ecology

Reproductive strategies are a set of attributes involved in an organism's maximizing its reproductive success. Theoretical and experimental studies of reproductive strategies reveal why various reproductive patterns have evolved.

The concept of reproductive strategies is closely related to that of natural selection. Natural selection results in the individuals within a population, under a given set of environmental circumstances, being more likely to pass on their genes to future generations. By this process, the gene pool (genetic makeup) of the population is altered over time. An organism's fitness can be assessed by evaluating two key characteristics: survival and reproductive success. The organism's reproductive strategy, then, is the blend of traits enabling it to have the highest overall reproductive success. Application of the term "reproductive strategy" has also been extended to describe patterns beyond individual organisms: populations, species, and even entire groups of similar species, such as carnivorous mammals.

Examination of reproductive strategies is part of the larger study of life-history evolution, which attempts to understand why a given set of basic traits has evolved. These traits include not only those pertaining to reproduction but also those such as body size and longevity. To consider a reproductive strategy appropriately, one must view it within the context of the organism's overall life history, precisely because these traits (particularly body size) often affect reproductive traits. One should also evaluate the role that the organism's ancestry plays in these processes. A species' evolutionary history can have a profound effect on its current attributes.

Reproductive Traits and Behaviors

A reproductive strategy consists of a collection of basic reproductive traits, including litter, or "clutch," size (the number of offspring produced per birth), the number of litters per year, the number of litters in a lifetime, and the time between litters, gestation, or pregnancy length. The age of the mother's first pregnancy is also a consideration. Another trait is the degree of development of the young at birth. In different species, mothers put varying levels of time and energy into the production of either relatively immature, or altricial, offspring or offspring that are well developed, or precocial.

Reproductive strategies also consist of behavioral elements, such as the mating system and the amount of parental care. Mating systems include monogamy (in which one male is mated to one female) and polygamy (in which an individual of one sex is mated to more than one from the other). The type of polygamy when one male mates with several females is called polygyny; the reverse is known as polyandry.

Finally, physiological events such as those involved in ovulation (what happens when the egg or eggs are shed from the ovary) may also be used to characterize a reproductive strategy. Some mammals are spontaneous ovulators. Females shed their eggs during the reproductive cycle without any physical stimulation. Other mammalian species are induced ovulators—a female ovulates only after being physically stimulated by a male during copulation. These patterns, induced and spontaneous ovulation, may be regarded as alternate reproductive strategies, each enabling a type of species to reproduce successfully under certain conditions.

The overall effectiveness of a reproductive strategy is important to consider with respect to the relative success of the offspring (even those in future generations) in leaving their own descendants. A sound reproductive strategy results in increased fitness. An organism's fitness as it affects the population's gene pool may not be adequately assessed until several generations have passed.

The r and K Selection Model

The model of r and K selection is the most widely cited description of how certain reproductive traits are most effective under certain environmental conditions. To appreciate this model, an understanding of elementary population dynamics is needed. At the early stages of a population's growth, the rate of addition of new individuals (designated r) tends to be slow. After a sufficient number of individuals is reached, the growth rate can increase sharply, resulting in a boom phase. In most environments, however, unrestrained growth cannot continue indefinitely. Critical resources—food, water, and protective cover—become more scarce as the environment's carrying capacity (K) is approached. Carrying capacity is the maximal population size an area can support. When the population approaches this level, growth rate slows as individuals now have fewer resources to convert into the production of new offspring.

This pattern is defined as density-dependent population growth—the density or number of individuals per area that influences its growth. This description of population dynamics is also referred to as logistic growth and was conceived by the Belgian mathematician Pierre-François Verhulst in the early nineteenth century. It has successfully described population growth in many species.

The r and K selection model was presented by Robert H. MacArthur and Edward O. Wilson in their influential book, *The Theory of Island Biogeography* (1967). They argued that in the early phase of a population's growth, individuals should evolve traits associated with high reproductive output. This enables them to take advantage of the relatively plentiful supply of food. The evolution of such traits is called r selection, after the high population growth rates occurring during this phase. They also suggested that, as the carrying capacity was approached, individuals would be selected that could adjust their lives to the now reduced circumstances. This process is called K selection. Such individuals should be more efficient in the conversion of food into offspring, producing fewer young than those living in the population's early phase. In a sense, a shift from productive to efficient individuals occurs as the population grows.

Other biologists, most notably Eric Pianka, have extended this concept of r and K selection to entire species rather than only to individuals at different stages of a population's growth. Highly variable or unpredictable climates commonly create situations in which population size is first diminished but then grows rapidly. Species commonly occurring in such environments are referred to as r strategists. Those living in more constant,

Male seals can successfully defend areas containing from eighty to a hundred females from other males. Very dense clusters of females, however, attract too many males for one male to monopolize. When this happens, the largest male typically dominates the rest and maintains disproportionate access to females. (PhotoDisc)

relatively predictable climates are less likely to go through such an explosive growth phase. These species are considered to be K strategists. According to this scheme, an r strategist is characterized by small body size, rapid development, high rate of population increase, early age of first reproduction, a single or few reproductive events, and many small offspring. The K strategist has the opposite qualities—large size, slower development, delayed age of first reproduction, repeated reproduction, and fewer, larger offspring.

Various combinations of r and K traits may occur in a species, and few are entirely r- or K-selected. Populations of the same species commonly occupy different habitats during their lives or across their geographic ranges. An organism might thus shift strategies in response to environmental changes—they may, however, be constrained by their phylogeny or ancestry in the degree to which their strategies are flexible.

Criticism of the Model
Because the r and K model of reproductive strategies seems to explain patterns observed in nature, it has become widely accepted. It has also met with considerable criticism. Charges against it include arguments that the logistic population-growth model (on which the r and K strategies model is based) is too simplistic. Another is that cases of r and K selection have not been adequately tested. Ecologist Mark Boyce has persuasively argued that for the r and K model to be most useful it must be viewed as a model of how population density affects life-history traits. Within this framework, also called density-dependent natural selection, the concept of r and K selection remains true to the one that MacArthur and Wilson originally proposed. Boyce suggests that the ability of r and K selection to explain reproductive strategies will have the best chance of being realized when approached in this fashion.

In addition to the r and K model, there are many other ways of describing reproductive strategies. For example, some species, such as the Chinook salmon, are semelparous: They reproduce only once before dying. The alternate is to be iteroparous—in which an organism experiences two or more reproductive events over its life span. If juvenile death rates are high, an individual might be better off reproducing on several occasions rather than only once. (This reproductive strategy is referred to as "bet-hedging.") Finally, it has also been useful to evaluate reproductive strategies based on the proportion of energy that goes into reproduction relative to that devoted to all other body functions. This mode of analysis addresses such considerations as reproductive effort and resource allocation.

Studying Reproductive Strategies

Initially, one who studies the reproductive strategy of an organism should attempt to characterize its reproduction fully. The sample examined must be representative of the population under consideration—it should account for the variability of the traits being measured. Studies can involve any of several approaches. Short-term laboratory studies can uncover some hard-to-observe features, but there is no substitute for long-term field research. By studying an organism's reproduction in nature, a biologist has the best chance of determining how its reproduction is shaped by an environment. If the research is performed over several seasons or years, patterns of variability can be better understood. This is important in determining how the physical environment influences reproductive traits.

After data have been systematically collected, it might then be possible to characterize a reproductive strategy. Imagine that a mouse population becomes established in a previously uninhabited area and that the population has a high reproductive rate (it produces large litters). The young develop quickly and produce many young themselves. Because of this combination of circumstances, one might consider the reproductive strategy to be r-selected since the population has a high reproductive output in an unexploited area. Though the concept of r and K strategies is problematic, it still is common to typify a strategy as r- or K-selected based upon this approach.

Because a reproductive strategy needs to be seen as part of an organism's overall life history, however, other things should be measured to understand it fully. These may include the life span and population attributes such as survival patterns. Values should be taken for different age groups to characterize the population's strategy. Correlational analysis is a statistical procedure that is used to evaluate reproductive strategies. Through such a methodology, one assesses the degree of association between two variables or factors. This may involve relationships between two reproductive variables or between a reproductive and an environmental variable—for example, to determine if there is a significant correlation between litter size and decreasing body size in mammals. If one were found to occur, the conclusion that smaller species typically have larger litters might be drawn, which is, in fact, true. Such an analysis enables the characterization of a change in reproductive strategy based on body size. Simply establishing a correlation does not prove that causation has occurred—it does not automatically mean that one factor is responsible for the expression of the other.

Multivariate statistical procedures are also used to analyze reproductive strategies. These allow the determination of how groups of reproduc-

tive traits are associated and of how they can be explained by several factors. One might determine that a certain bird species produces its greatest
number of young, and that the young grow most rapidly, at northern locations having high snow levels. Such an approach is often needed in dealing
with reproductive strategies—a combination of traits typically requires explanation.

Reproduction and Survivial

The characterization of an organism's reproductive strategy involves more
than an understanding of reproductive traits. There is a successful process
by which offspring are produced, and reproductive success is one of the
two principal measures of fitness—the other is survival. Because a successful reproductive strategy ultimately results in high fitness, any discussion
of these strategies bears directly on issues of natural selection and evolution.

An organism's reproductive strategy represents perhaps the most significant way the organism is adapted to its environment. A successful reproductive strategy represents a successful mode of passing genes on to
the next generation, so traits associated with a reproductive strategy are
under intense natural selection pressure. If environmental conditions
change, the original strategy may no longer be as successful. To the extent
that an organism can shift its reproductive strategy as circumstances
change, its genes will persist.

The study of reproductive strategies has helped scientists understand
why certain modes of reproduction occur, based upon observations of a
species itself and of its environment. An understanding of reproductive
strategies may also be of some practical use. An organism's reproduction
directly influences its population dynamics.

If an animal has small litters and is at an early age at first reproduction,
its population should grow at a concomitantly high rate. These and other
components of reproduction may strongly affect a species' population
growth. A knowledge of how reproduction influences population dynamics can be important in wildlife management activities, which can range
from strict preservation efforts to overseeing trophy hunting.

Samuel I. Zeveloff

See also: Adaptive radiation; Biodiversity; Biogeography; Clines, hybrid
zones, and introgression; Convergence and divergence; Demographics;
Displays; Ethology; Extinctions and evolutionary explosions; Gene flow;
Genetic diversity; Genetic drift; Human population growth; Insect societies; Natural selection; Nonrandom mating, genetic drift, and mutation;

Population analysis; Population fluctuations; Population genetics; Population growth; Punctuated equilibrium vs. gradualism; Speciation; Territoriality and aggression.

Sources for Further Study

Austin, C. R., and R. V. Short, eds. *The Evolution of Reproduction*. New York: Cambridge University Press, 1976.

Boyce, Mark S. "Restitution of r- and K-Selection as a Model of Density-Dependent Natural Selection." *Annual Review of Ecology and Systematics* 15 (1984): 427-447.

Clutton-Brock, T. H., F. E. Guiness, and S. D. Albon. *Red Deer: Behavior and Ecology of Two Sexes*. Chicago: University of Chicago Press, 1982.

Ferraris, Joan D., and Stephen R. Palumbi, eds. *Molecular Zoology: Advances, Strategies, and Protocols*. New York: Wiley-Liss, 1996.

MacArthur, Robert H., and Edward O. Wilson. *The Theory of Island Biogeography*. Princeton, N.J.: Princeton University Press, 1967.

Pianka, Eric R. *Evolutionary Ecology*. 6th ed. New York: Harper & Row, 2000.

Wrangham, Richard W., W. C. McGrew, Frans B. M. De Waal, and Paul G. Heltne, eds. *Chimpanzee Cultures*. Cambridge, Mass.: Harvard University Press, 1996.

RESTORATION ECOLOGY

Type of ecology: Restoration and conservation ecology

Restoration ecology is concerned with converting ecosystems that have been modified or degraded by human activity to a state approximating their original condition.

Federal laws often dictate ecological restoration following strip mining, construction, and other activities that alter the landscape. As a part of the management of natural areas that have been disturbed to some degree, several options are available. One is to do nothing but protect the property, allowing nature to take its course. In the absence of further disturbances, one would expect the area to undergo the process of ecological succession. Theoretically, an ecosystem similar to that typical of the region, and including an array of organisms, would be expected to return.

One might, therefore, ask why ecological restoration is mandated. For one thing, succession is often a process requiring long periods of time. As an example, the return of a forest following the destruction of the trees and the removal of the soil would require more than one century. Also, ecosystems resulting from succession may be lacking in species typical of the region. This is true when succession is initiated in an area where many exotic (alien) species are present or where certain native species have been eliminated. Succession can produce a new ecosystem with a biodiversity comparable to the original one only if there is a local source of colonizing animals and seeds of native plants. Also, satisfactory recovery by succession is unlikely if the soil has been heavily polluted by heavy metals or other substances caused by industrial land use.

The Restoration Process

Once it has been decided that a given ecosystem is to be restored, success requires that a plan be designed and followed. Although the specifics may vary greatly, all restoration projects should follow five basic steps: Envision the end result, consult relevant literature and solicit the advice of specialists, remove or mitigate any current disturbances to the site, rehabilitate the physical habitat, and restore indigenous plants and animals.

Much can be learned from restoration projects that have been conducted in various parts of the world involving a wide variety of ecosystems. A classic ecological restoration of a prairie was conducted in Wisconsin beginning in the 1930's. Because most North American prairies have

Restoration ecology enlists human intervention to return habitats and ecosystems to their former state, particularly when natural processes such as ecological succession would take far longer than the original destruction of the community. Such destruction is often the result of human activity such as clear-cutting, strip-mining, large-scale agriculture, or other development. (PhotoDisc)

been converted to agricultural uses, many opportunities exist for prairie restoration. In such projects it is often necessary to eliminate exotic plants by mechanical means or by application of herbicides. Native prairie grasses and forbs can be established by transplantation or from seed. It may also be necessary to introduce native fauna from nearby areas. Periodic prescribed burning is often necessary to simulate natural fires common in prairies.

After decades of loss of wetlands in the United States, governmental policy is now "no net loss." Thus, when a wetland is destroyed by development, it is required that a new wetland be created as compensation. Before introducing native biota, it is necessary to alter the hydrology of the new site.

Thomas E. Hemmerly

See also: Biological invasions; Conservation biology; Deforestation; Endangered animal species; Endangered plant species; Erosion and erosion control; Forest management; Grazing and overgrazing; Integrated pest management; Invasive plants; Landscape ecology; Multiple-use approach;

Old-growth forests; Reforestation; Succession; Sustainable development; Urban and suburban wildlife; Waste management; Wildlife management; Zoos.

Sources for Further Study

Harper, David. *Eutrophication of Freshwaters: Principles, Problems, and Restoration*. London: Chapman & Hall, 1992.
Hey, Donald L., and Nancy S. Philippi. *A Case for Wetland Restoration*. New York: Wiley, 1999.

SAVANNAS AND DECIDUOUS TROPICAL FORESTS

Types of ecology: Biomes; Ecosystem ecology

Savannas are areas of continuous grass or sedge cover beneath trees that range from scattered, twisted, and gnarled individuals to open woodlands. Deciduous tropical forests have continuous to open forest cover and undergo a leafless period during a seasonally lengthy dry season.

Where the annual rainfall in tropical regions is less than 2,000 millimeters and three to six months out of the year are dry, savannas and deciduous forests are common. Deciduous tropical forests often occur where the annual rainfall is less than that of savannas. Together, the two biomes are referred to here as the dry tropical biome.

A pronounced pattern of seasonally wet and dry periods is the most important factor affecting the distribution of these types of plant cover. Higher soil fertility favors forest over grasses and savanna such as in the cerrado of Brazil, which occurs only on certain geological formations and low-nutrient soils. Fire has been a dominant feature of these biomes, and human influences—fires, agriculture, and grazing of animals—have interacted with climate to produce a varied landscape.

The dry tropical biome is most geographically widespread on the continents of Africa, South America, and Australia, with smaller enclaves in Asia. The world's largest expanses of dry forest—the Brachystegia woodland across Central Africa, the cerrado (savanna) and caatinga (dry forest) of the Amazon basin, and much of interior Australia—are notable examples. "Elephant grass savanna," with tall grasses up to 4 meters tall and scattered trees, occurs exclusively in Africa. In the West Indies, dry forest occurs in rain-shadow zones on the leeward sides of islands affected by the tradewinds.

Plant Adaptations and Diversity of Life-Forms

As the rainfall decreases below 2,000 millimeters, and especially below 1,000 millimeters, the height of the forest decreases and the proportion of trees that are deciduous increases. In the dry tropics, leaf fall occurs in response to drought, and therefore the lengthy dry season becomes a selective pressure to which plants have adapted. Tree leaves tend to be compound, with small leaflets that help plants exchange heat with their surroundings better than large, simple leaves; rates of leaf respiration and

586

transpiration are thereby reduced. Sclerophyllous leaves are common, aiding in moisture retention, and the drier, more open woodlands may have cacti or other succulents.

The dry forest is far less species-rich than the rain forest, but the diversity of life-forms and the proportion of endemics are greater. For example, dry forests may contain xerophytic (dry-adapted) evergreens, either obligatively or facultatively deciduous trees, trees with photosynthetic bark, plants that use the Crassulacean acid metabolism (CAM) photosynthesis as well as C_3 and C_4 dicots, grasses, bromeliads, lianas, epiphytes, and hemiparasites. Trees from *Fabaceae* (the legume family) are the most well-represented family among trees.

Dry forests contain a higher proportion of wind-dispersed species than wetter forests, and many trees will have their flowering and fruiting controlled by the duration and intensity of the dry season. Synchronous flowering within and among species is common, and many produce seed during the dry season. Flowers are often conspicuous and visited by specialized pollinators such as hawkmoths, bats, and bees.

It is incorrect to generalize about savannas and dry tropical forests because, although they both occur in the drier tropics, the two vegetation

Savannas are landscapes of dense grass and scattered trees, such as these yellow fever trees growing in Nakura National Park in Kenya. Common on the continent of Africa, savannas are also found in India, Australia, and the northern part of South America. (Corbis)

types occur in different habitats and are adapted differently to their respective environments.

Trees of the cerrado in northeast Brazil are deeply rooted, tap groundwater, and have high rates of transpiration. Drought here is atmospheric, as water is always available below two meters of soil depth. The deciduous caatinga of central Brazil, however, receives only 500 millimeters of rain yearly, and transpiration of trees is low. Here, trees suffer significant water deficits during the long, dry season, are truly xerophytic, and exhibit classic adaptations to drought.

Trees of the cerrado have a number of adaptations that confer resistance to fire. These include a thick, corky bark, the ability to form adventitious roots from buds on roots following the burning of the stem, and the cryptophyte or hemicryptophyte life-form (cryptophytes produce buds underground). Many herbaceous species are induced to flower by fire.

Human Impacts and Conservation

Fires have occurred in the Brazilian cerrado for thousands of years, based on carbon 14 dating of charcoal fragments. Fire is thus an environmental factor to which the vegetation has become adapted. Yet, the human influence has been to increase the incidence of fire. The cerrado has changed as a result to a more open form of plant cover with fewer trees and shrubs. In addition, timber extraction, charcoal production, and ranching have altered the savanna landscape. The ability of belowground organs to survive such types of disturbance has increased the ability of the cerrado to persist. Yet it is estimated that 50 percent of the cerrado has been destroyed, much of this the result of clearing for agriculture since the 1960's.

Because of better soils and fewer pests, humans in tropical areas of Central America have mostly chosen the dry and moist life zones as places to live rather than the wetter rain-forest zones. As a result, dry forest ecosystems have been subject to massive disturbance. Today, only a small fraction of the original dry forest remains. Fire has been used as a means of clearing the forest for farming, but, unlike the savanna, the dry forest is not adapted to fire. At Guanacaste, Costa Rica, a well-known tropical conservation project, restoration of dry forest is dependent on controlling annual fires set by farmers and ranchers and supporting the return of forest vegetation to dry areas. In Africa, large areas of dry forest are burned annually by farmers, and areas of dense, dry forest have been converted to more open forest or even savanna. Sustainable land-use systems are urgently needed for dry tropical regions.

Allan P. Drew

See also: Biomes: determinants; Biomes: types; Forests; Grasslands and prairies; Habitats and biomes; Rangeland; Reforestation.

Sources for Further Study

Allen, W. *Green Phoenix: Restoring the Tropical Forests of Guanacaste, Costa Rica*. New York: Oxford University Press, 2001.

Bullock, S. H., H. A. Mooney, and E. Medina. *Seasonally Dry Tropical Forests*. New York: Cambridge University Press, 1995.

Rizzini, C. T., A. F. C. Filho, and A. Houaiss. *Brazilian Ecosystems*. Rio de Janeiro: ENGE-RIO, Index Editora, 1988.

Silver, Donald M. *African Savanna*. New York: McGraw-Hill, 1997.

SLASH-AND-BURN AGRICULTURE

Types of ecology: Agricultural ecology; Ecotoxicology

Slash-and-burn agriculture, also called swidden agriculture, is a practice in which forestland is cleared and burned for use in crop and livestock production. While yields are high during the first few years, they rapidly decline in subsequent years, leading to further clearing of nearby forestland.

Slash-and-burn agriculture has been practiced for many centuries among people living in tropical rain forests. Initially, this farming system involved small populations. Therefore, land could be allowed to lie fallow (unplanted) for many years, leading to the full regeneration of the secondary forests and hence a restoration of the ecosystems. During the second half of the twentieth century, however, several factors led to drastically reduced fallow periods. In some places such fallow systems are no longer in existence, resulting in the transformation of forests into shrub and grasslands, negative effects on agricultural productivity for small farmers, and disastrous consequences to the environment.

Among the factors that have been responsible for reduced or nonexistent fallow periods are increased population in the tropics, increased demand for wood-based energy, and, perhaps most important, the increased worldwide demand for tropical commodities during the 1980's and 1990's, especially for products such as palm oil and natural rubber. These last two factors have helped industrialize slash-and-burn agriculture, which was practiced for centuries mainly by small farmers. Ordinarily, small farmers are able to control their fires so that they are similar to a small forest fire triggered by lightning in the northwestern or southeastern United States. However, the continued reduction in fallow periods, coupled with increased burning by subsistence farmers and large agribusiness, especially in Asia and Latin America, is resulting in increased environmental concern.

While slash-and-burn agriculture seldom takes place in temperate regions, some agricultural burning occurs in the Pacific Northwest of the United States, where it is estimated that three thousand to five thousand agricultural fires are set each year in Washington State alone. These fires also create problems for human health and the environment.

Habitat Fragmentation

One of the most easily recognized results of slash-and-burn agriculture is habitat fragmentation, which leads to a significant loss of the vegetation needed for the maintenance of effective gaseous exchange in tropical regions and throughout the world. For every acre of land lost to slash-and-burn agriculture, 10 to 15 acres (4 to 6 hectares) of land are fragmented, resulting in the loss of habitat for wildlife, plant species, and innumerable macro- and microorganisms yet to be identified. This also creates problems for management and wildlife conservation efforts in parts of the world with little or no resources to feed their large populations. Fragmentation has also led to intensive discussions on global warming. While slash-and-burn agriculture by itself is not completely responsible for global warming, the industrialization of the process could make it a significant component of the problem, as more and more vegetation is fragmented.

Human Health

The impact of slash-and-burn agriculture on human health and the environment is best exemplified by the 1997 Asian fires that resulted from such practices. Monsoon rains normally extinguish the fires set by farmers, but a strong El Niño weather phenomenon delayed the expected rains, and the fires burned out of control for months. Thick smoke caused severe health problems. It is estimated that more than 20 million people in Indonesia alone were treated for asthma, bronchitis, emphysema, and eye, skin, and cardiovascular problems as a result of the fires. Similar problems have been reported for smaller agricultural fires.

Three major problems are associated with air pollution: particulate matter, pollutant gases, and volatile organic compounds. Particulate compounds of 10 microns or smaller that are inhaled become attached to the alveoli and other blood cells, resulting in severe illness. Studies by the U.S. Environmental Protection Agency (EPA) and the University of Washington indicate that death rates associated with respiratory illnesses increase when fine particulate air pollution increases. Meanwhile, pollutant gases such as carbon monoxide, nitric oxide, nitrogen dioxide, and sulfur dioxide become respiratory irritants when they combine with vapor to form acid rain or fog. Until the Asian fires, air pollutants stemming from the small fires of slash-and-burn agriculture that occur every planting season often went unnoticed. Thus, millions of people in the tropics experience environmental health problems because of slash-and-burn agriculture that are never reported.

Soil and Water Quality

The loss of vegetation that follows slash-and-burn agriculture causes an increased level of soil erosion. The soils of the humid tropics create a hard pan underneath a thick layer of organic matter. Therefore, upon the re-

A bean farmer in Nicaragua walks away from cut vegetation, or slash, he has set on fire. The land is burned in preparation for planting, and ashes from the burned trees fertilize the soil. (AP/Wide World Photos)

moval of vegetation cover, huge areas of land become exposed to the torrential rainfalls that occur in these regions. The result is severe soil erosion. As evidenced by the impact of Hurricane Mitch on Honduras during 1998, these exposed lands can give rise to large mudslides that can lead to significant loss of life. While slash-and-burn agriculture may not be the ultimate cause for sudden mud slides, it does predispose these lands to erosional problems.

Associated with erosion is the impact of slash-and-burn agriculture on water quality. As erosion continues, sedimentation of streams increases. This sedimentation affects stream flow and freshwater discharge for catchment-area populations. Mixed with the sediment are minerals such as phosphorus and nitrogen-related compounds that enhance algal growth in streams and estuaries, which depletes the supply of oxygen that aquatic organisms require to survive. Although fertility is initially increased on noneroded soils, nutrient deposition and migration into drinking water supplies continues to increase.

Controlling Slash-and-Burn Agriculture
Given the fact that slash-and-burn agriculture has significant effects on the environment not only in regions where it is the mainstay of the agricultural

systems but also in other regions of the world, it has become necessary to explore different approaches to controlling this form of agriculture. However, slash-and-burn agriculture has evolved into a sociocultural livelihood; therefore, recommendations must be consistent with the way of life of a people who have minimal resources for extensive agricultural systems.

Among the alternatives are new agroecosystems such as agroforestry systems and sustainable agricultural systems that do not rely so much on the slashing and burning of forestlands. These systems allow for the cultivation of agronomic crops and livestock within forest ecosystems. This protects soils from being eroded. Another possibility is the education of small rural farmers, absentee landlords, and big agribusiness concerns in developing countries to understand the environmental impact of slash-and-burn agriculture. While small rural farmers may not have the resources for renovating utilized forestlands, big business can organize ecosystems restoration, as has been done in many developed nations of the world.

Oghenekome U. Onokpise

See also: Biopesticides; Deforestation; Erosion and erosion control; Forest management; Forests; Global warming; Grazing and overgrazing; Multiple-use approach; Pesticides; Rangeland; Savannas and deciduous tropical forests; Sustainable development.

Sources for Further Study

Jordan, C. F. *An Amazonian Rain Forest: The Structure and Function of a Nutrient Stressed Ecosystem and the Impact of Slash and Burn Agriculture.* Boca Raton, Fla.: CRC Press, 1990.

Simons, L. M., and M. Yamashita. "Indonesia's Plague of Fire." *National Geographic* 194 (August, 1998): 100-120.

Terborgh, J. *Diversity and the Tropical Rain Forest.* New York: Scientific American Library, 1992.

Youth, Howard. "Green Awakening in a Poor Country." *World Watch* 11, no. 5 (September/October, 1998): 28-37.

SOIL

Types of ecology: Agricultural ecology; Ecosystem ecology; Soil ecology

Soils are complex chemical factories. Regardless of the type of soil—and twelve types of soil are identified by the U.S. Department of Agriculture—chemical processes such as plant growth, organic decay, mineral weathering, and water purification, as well as living organisms, constitute the ecosystem commonly referred to as soil.

Soil chemistry has been studied as long as there has been sustainable agriculture. Although they did not recognize it as such, those first successful farmers who plowed under plant stalks, cover crops, or animal wastes were actively managing the soil chemistry of their fields. These early farmers knew that to have productive farms in one location season after season, they had to return something to the soil.

It is now understood that soil chemistry is a complex of chemical and biochemical reactions. The most obvious result of this complex of reactions is that some soils are very fertile whereas other soils are not. Soil itself is a unique environment because all the "spheres"—the atmosphere, hydrosphere, geosphere, and biosphere—are intimately mixed there. For this reason, soil and soil chemistry are extremely important.

Rock Weathering

Soil chemistry begins with rock weathering. The minerals composing a rock exposed at the earth's surface are continually bathed in a shower of acid rain—not necessarily polluted rainwater but naturally occurring acid rain. Each rain droplet forming in the atmosphere absorbs a small amount of carbon dioxide gas. Some of the dissolved carbon dioxide reacts with the water to form a dilute solution of carbonic acid. A more concentrated solution of carbonic acid is found in any bottle of sparkling water.

Most of the common rock-forming minerals, such as feldspar, will react slowly with rainwater. Some of the chemical elements of the mineral, such as sodium, potassium, calcium, and magnesium, are very soluble in rainwater and are carried away with the water as it moves over the rock surface. Other chemical elements of the mineral, such as aluminum, silicon, and iron, are much less soluble. Some of these elements are dissolved in the water and carried away; most, however, remain near the original weathering, where they recombine into new, more resistant minerals. Many of the new minerals are of a type called clays.

Clay minerals tend to be very small crystals composed of layers of aluminum and silicon. Between the layers of aluminum and silicon atoms are positively charged ions (cations) of sodium, potassium, calcium, and magnesium. The cations hold the layers of some clays together by electrostatic attraction. In most cases, the interlayer cations are not held very tightly. They can migrate out of the clay and into the water surrounding the clay mineral, to be replaced by another cation from the soil solution. This phenomenon is called cation exchange.

The weathering reactions between rainwater and rock minerals produce a thin mantle of clay mineral soil. The depth to fresh, unweathered rock is not great at first, but rainwater continues to fall, percolating through the thin soil and reacting with fresh rock minerals. In this way, the weather-

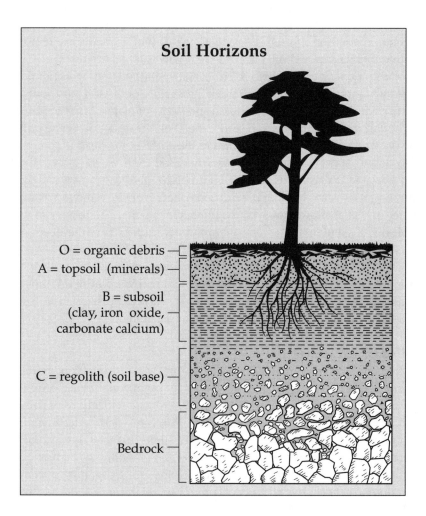

Soil Horizons

O = organic debris
A = topsoil (minerals)
B = subsoil (clay, iron oxide, carbonate calcium)
C = regolith (soil base)
Bedrock

ing front (the line between weathered minerals and fresh rock) penetrates farther into the rock, and the overlying soil gets thicker.

Biological Processes

Throughout the weathering process, biological processes contribute to the pace of soil formation. In the very early stages, lichens and fungi are attached to what appear to be bare rock surfaces. In reality, they are using their own acids to "digest" the rock minerals. They absorb the elements of the mineral they need, and the remainder is left to form soil minerals. As the soil gets thicker, larger plants and animals begin to colonize. Large plants send roots down into the soil looking for water and nutrients. Some of the necessary nutrients, such as potassium, are available as exchangeable cations on soil clays or in the form of deeper, unweathered minerals. In either case, the plant obtains the nutrients by using its own weathering reaction carried on through its roots. The nutrient elements are removed from minerals and become part of the growing plant's tissue.

Without a way to replenish the nutrients in the soil, the uptake of nutrients by plants will eventually deplete the fertility of the soil. Nutrients are returned to the soil through the death and decay of plants. Microorganisms in the soil, such as bacteria and fungi, speed up the decay. Since the bulk of the decaying plant material is found at the surface (the dead plant's roots also decay), most of the nutrients are released to the surface layer of the soil. Some of the nutrients are carried down to roots deep in the soil by infiltrating rainwater. Most of the nutrients, however, are removed from the water by the shallow root systems of smaller plants. The deeper roots of typically large plants can mine the untapped nutrients at the deep, relatively unweathered soil-rock boundary.

The soil and its soil chemistry are now well established, with plants growing on the surface and their roots reaching toward mineral nutrients at depth. Water is flowing through the soil, carrying dissolved nutrients and the soluble by-products of weathering reactions.

Soil as an Ecosystem

Not to be forgotten in this mix are the microbes, insects, nematodes, worms, and other organisms that, along with fungi and plants, occupy the soil ecosystem. There are at least twelve different classifications of soil recognized by the U.S. Department of Agriculture, and these reflect the many communities that occupy various soil ecosystems. In fact, when ecologists discuss soil as an ecosystem, they are referring not only to the biotic and abiotic components of soil itself but also to the living organisms, from all kingdoms, that occupy, partially occupy, or travel through soil. Disruption to soil by herbicides, polluted water, solid waste, mechanical erosion, and other factors therefore has a reverberating impact not only on the minerals and chemical compounds that form soil but also on all members of the soil ecosystem.

Soil Chemistry

Soil ecologists and soil chemists are concerned not only with the composition of soil and soil water but also with how that composition changes as the water interacts with the atmosphere, minerals, plants, fungi, animals, and mechanical forces at work on it. Soil and its chemistry can be studied in its natural environment, or samples can be brought into the laboratory for testing. Some tests have been standardized and are best conducted in the laboratory so that they can be compared with the results of other researchers. Most of the standardized tests, such as measures of the soil's acidity and cation-exchange capacity, are related to measures of the soil's fertility and its overall suitability for plant growth. These tests measure average values for a soil sample because large original samples are dried and thoroughly mixed before smaller samples are taken for the specific test.

Increasingly, soil chemists are looking for ways to study the fine details of soil chemical processes. They know, for example, that soil water chemistry changes as the water percolates through succeeding layers of the soil. The water flowing through the soil during a rainstorm has a different chemical composition from that of water clinging to soil particles, at the same depth, several days later. Finally, during a rainstorm, the water flowing through large cracks in the soil has a chemical composition different from that of the same rainwater flowing through the tiny spaces between soil particles.

Sampling Techniques

Soil chemists use several sampling techniques to collect the different types of soil water. During a rainstorm, water flows under the influence of gravity. After digging a trench in the area of interest, researchers push several sheets of metal or plastic, called pan lysimeters, into the wall of the trench at specified depths below the surface. The pans have a very shallow V shape. Soil water flowing through the soil collects in the pan, flows toward the bottom of the V, and flows out of the pan into a collection bottle. Comparing the chemical compositions of rainwater that has passed through different thicknesses of soils (marked by the depth of each pan) allows the soil chemist to identify specific soil reactions with specific depths.

After the soil water stops flowing, water is still trapped in the soil. The soil water clings to soil particles and is said to be held by tension. Tension water can spend a long time in the soil between rainstorms. During that

The Twelve Soil Orders in the U.S. Classification System

Soil Order	Features
Alfisols	Soils in humid and subhumid climates with precipitation from 500 to 1,300 millimeters (20 to 50 inches), frequently under forest vegetation. Clay accumulation in the B horizon and available water most of the growing season. Slightly to moderately acid soils.
Andisols	Soils with greater than 60 percent volcanic ash, cinders, pumice, and basalt. They have a dark A horizon as well as high absoption and immobilization of phosphorus and very high cation exchange capacity.
Aridisols	Aridisols exist in dry climates. Some have horizons of lime or gypsum accumulations, salty layers, and A and slight B horizon development.
Entisols	Soils with no profile development except a shallow A horizon. Many recent river floodplains, volcanic ash deposits, severely eroded areas, and sand are entisols.
Gelisols	Soils that commonly have a dark organic surface layer and mineral layers underlain by permafrost, which forms a barrier to downward movement of soil solution. Common in tundra regions of Alaska. Alternate thawing and freezing of ice layers results in special features in the soil; slow decomposition of the organic matter due to cold temperatures results in a peat layer at the surface in many gelisols.
Histosols	Organic soils of variable depths of accumulated plant remains in bogs, marshes, and swamps.
Inceptisols	Soils found in humid climates that have weak to moderate horizon development. Horizon development may have been delayed because of cold climate or waterlogging.
Mollisols	Mostly grassland soils, but with some broadleaf forest-covered soils with relatively deep, dark A horizons, a possible B horizon, and lime accumulation.
Oxisols	Excessively weathered soils. Oxisols are over 3 meters (10 feet) deep, have low fertility, have dominantly iron and aluminum oxide clays, and are acid. Oxisols are found in tropical and subtropical climates.
Spodosols	Sandy leached soils of the cool coniferous forests, usually with an organic or O horizon and a strongly acidic profile. The distinguishing feature of spodosols is a B horizon with accumulated organic matter plus iron and aluminum oxides.
Ultisols	Strongly acid and severely weathered soils of tropical and subtropical climates. They have clay accumulation in the B horizon.
Vertisols	Soils with a high clay content that swell when wet and crack when dry. Vertisols exist in temperate and tropical climates with distinct dry and wet seasons. Usually vertisols have only a deep self-mixing A horizon. When the topsoil is dry, it falls into the cracks, mixing the soil to the depth of the cracks.

time, it reacts with soil mineral grains and soil microorganisms. Tension water is sampled by placing another type of lysimeter, a tension lysimeter, into the soil at a known depth. A tension lysimeter is like the nozzle of a vaccum cleaner with a filter over the opening. Soil chemists actually vacuum the tension water out of the soil and to the surface for analysis.

Determining Isotopic Composition

Nonradioactive, stable isotopes of common elements are being used more often by soil chemists to trace both the movement of water through the soil and the chemical reactions that change the composition of the water. Trace stable isotopes behave chemically just the way their more common counterparts do. For example, deuterium, an isotope of hydrogen, substitutes for hydrogen in the water molecule and allows the soil chemist to follow the water's movements. Similarly, carbon 13 and nitrogen 15 are relatively rare isotopes of common elements that happen to be biologically important. Using these isotopes, soil chemists can study the influences of soil organisms on the composition of soil water. Depending on what the soil chemist is studying, the isotope may be added, or spiked, to the soil in the laboratory or in the field. Alternatively, naturally occurring concentrations of the isotope in rain or snowmelt may be used. Regardless, soil water samples are collected by one or more of the lysimeter methods, and their isotopic composition is determined.

The Soil Chemical Factory

The wonderful interactions of complex chemical and biochemical reactions that are soil chemistry are one indication of the uniqueness of planet Earth. Without the interaction of liquid water and the gases in the atmosphere, many of the nutrients necessary for life would remain locked up in rock minerals. Thanks to weathering reaction, the soil chemical factory started to produce nutrients, which resulted in the exploitation of the soil environment by millions of organisms. The processes involved in soil chemistry—from weathering reactions that turn rock into new soil to the recycling of plant nutrients through microbial decay—are vital to every human being. Without fertile soil, plants will not grow. Without plants as a source of oxygen and food, there would be no animal life.

Because of the complex chemical interrelationships that have developed in the soil environment, it may seem that nothing can disrupt the "factory" operation. As more is understood about soil chemistry and the ways in which humans stress the soil chemistry through their activities, it is apparent that the factory is fragile. Not only do humans rely on soil fertility for their very existence, but they also are taking advantage of soil chemical processes to help them survive their own past mistakes. Soil has been and continues to be used as a garbage filter. Garbage, whether solid or liquid, has been dumped on or buried in soil for ages. Natural chemical processes broke down the garbage into simpler forms and recycled the nutrients. When garbage began to contain toxic chemicals, those chemicals, when in small quantities, were either destroyed by soil bacteria or firmly attached to soil particles. The result is that water—percolating through garbage, on its way to the local groundwater, stream, or lake—does not carry with it as much contamination as one might expect. Soil chemistry has, so far, kept contaminated garbage from ruining drinking water. There are well-known cases, however, where the volume and composition of waste buried or spilled were such that the local soil chemistry was overwhelmed. In cases of large industrial spills, or when artificial chemicals are spilled or buried, the soil needs help to recover. The recovery efforts are usually very expensive but, faced with the possible permanent loss of large parts of the soil chemical factory, humankind cannot afford to neglect this aspect of the environment.

Richard W. Arnseth

See also: Acid deposition; Deforestation; Endangered plant species; Erosion and erosion control; Grasslands and prairies; Grazing and overgrazing; Multiple-use approach; Nutrient cycles; Pesticides; Rangeland; Reforestation; Slash-and-burn agriculture; Soil contamination.

Sources for Further Study

Berner, Elizabeth K., and Robert A. Berner. *The Global Water Cycle: Geochemistry and Environment*. Englewood Cliffs, N.J.: Prentice-Hall, 1987.

Bohn, Heinrich, B. L. McNeal, and G. A. O'Connor. *Soil Chemistry*. 2d ed. New York: Wiley, 1985.

Brill, Winston. "Agricultural Microbiology." *Scientific American* 245 (September, 1981): 198.

Evangelou, V. P. *Environmental Soil and Water Chemistry: Principles and Applications*. New York: Wiley, 1998.

Lloyd, G. B. *Don't Call It Dirt*. Ontario, Calif.: Bookworm Publishing, 1976.

McBride, Murray B. *Environmental Chemistry of Soils*. New York: Oxford University Press, 1994.

Millot, Georges. "Clay." *Scientific American* 240 (April, 1979): 108.

Sparks, Donald S., ed. *Soil Physical Chemistry*. 2d ed. Boca Raton, Fla.: CRC Press, 1999.

Tan, Kim Howard. *Principles of Soil Chemistry*. 3d ed. New York: M. Dekker, 1998.

SOIL CONTAMINATION

Types of ecology: Ecotoxicology; Restoration and conservation ecology;
Soil ecology

Soils contaminated with high concentrations of hazardous substances pose poten-
tial risks to human health and the earth's thin layer of productive soil.

Productive soil depends on bacteria, fungi, and other soil microbes to break down wastes and release and cycle nutrients that are essential to plants. Healthy soil is essential for growing enough food for the world's increasing population. Soil also serves as both a filter and a buffer between human activities and natural water resources, which ultimately serve as the primary source of drinking water. Soil that is contaminated may serve as a source of water pollution through leaching of contaminants into groundwater and through runoff into surface waters such as lakes, rivers, and streams.

Types of Contamination

Soils can become contaminated by many human activities, including fertilizer or pesticide application, direct discharge of pollutants at the soil surface, leaking of underground storage tanks or pipes, leaching from landfills, and atmospheric deposition. Additionally, soil contamination may be of natural origin. For example, soils with high concentrations of heavy metals can occur naturally because of their close proximity to metal ore deposits. Common contaminants include inorganic compounds such as nitrate and heavy metals (for example, lead, mercury, cadmium, arsenic, and chromium); volatile hydrocarbons found in fuels, such as benzene, toluene, ethylene, and xylene BTEX compounds; and chlorinated organic compounds such as polychlorinated biphenyls (PCBs) and pentachlorophenol (PCP).

Contaminants may also include substances that occur naturally but whose concentrations are elevated above normal levels. For example, nitrogen- and phosphorus-containing compounds are often added to agricultural lands as fertilizers. Since nitrogen and phosphorus are typically the limiting nutrients for plant and microbial growth, accumulation in the soil is usually not a concern. The real concern is the leaching and runoff of the nutrients into nearby water sources, which may lead to oxygen depletion of lakes as a result of the eutrophication encouraged by those nutrients. Furthermore, nitrate is a concern in drinking water because it poses a direct risk to human infants; it is associated with blue-baby syndrome.

Environmental Interactions

Contaminants may reside in the solid, liquid, and gaseous phases of the soil. Most will occupy all three phases but will favor one phase over the others. The physical and chemical properties of the contaminant and the soil will determine which phase the contaminant favors. The substance may preferentially adsorb to the solid phase, either the inorganic minerals or the organic matter. The attraction to the solid phase may be weak or strong. The contaminant may also volatize into the gaseous phase of the soil. If the contaminant is soluble in water, it will dwell mainly in the liquid-filled pores of the soil.

Contaminants may remain in the soil for years or make their way into the atmosphere or nearby water sources. Additionally, the compounds may be broken down or taken up by the biological component of the soil. This may include plants, bacteria, fungi, and other soil-dwelling microbes. The volatile compounds may slowly move from the gaseous phase of the soil into the atmosphere. The contaminants that are bound to the solid phase may remain intact or be carried off in runoff attached to soil particles and flow into surface waters. Compounds that favor the liquid phase, such as nitrate, will either move into surface waters or leach down into the groundwater.

Metals display a range of behaviors. Some bind strongly to the solid phase of the soil, while others easily dissolve and wind up in surface or groundwater. PCBs and similar compounds bind strongly to the solid surface and remain in the soil for years. These compounds can still pose a threat to waterways because, over long periods of time, they slowly dissolve from the solid phase into the water at trace quantities. Fuel components favor the gaseous phase but will bind to the solid phase and dissolve at trace quantities into the water. However, even trace quantities of some compounds can pose a serious ecological or health risk. When a contaminant causes a harmful effect, it is classified as a pollutant.

Treatments

There are two general approaches to cleaning up a contaminated soil site: treatment of the soil in place (in situ) or removal of the contaminated soil followed by treatment (non-in situ). In situ methods, which have the advantage of minimizing exposure pathways, include biodegradation, volatilization, leaching, vitrification (glassification), and isolation or containment. Non-in situ methods generate additional concerns about exposure during the process of transporting contaminated soil. Non-in situ options include thermal treatment (incineration), land treatment, chemical extraction, solidification or stabilization, excavation, and asphalt incorporation.

The choice of methodology will depend on the quantity and type of contaminants and on the nature of the soil. Some of these treatment technologies are still in the experimental phase.

John P. DiVincenzo

See also: Acid deposition; Biomagnification; Biopesticides; Erosion and erosion control; Food chains and webs; Geochemical cycles; Hydrologic cycle; Integrated pest management; Pesticides; Pollution effects; Soil; Waste management.

Sources for Further Study
Pierzynski, Gary M., J. Thomas Sims, and George F. Vance. *Soils and Environmental Quality.* Boca Raton, Fla.: Lewis, 1994.
Sparks, Donald L. "Soil Decontamination." In *Handbook of Hazardous Materials,* edited by Morton Corn. San Diego, Calif.: Academic Press, 1993.
Testa, Stephen M. *The Reuse and Recycling of Contaminated Soil.* Boca Raton, Fla.: Lewis, 1997.

SPECIATION

Types of ecology: Community ecology; Evolutionary ecology; Speciation

Processes whereby new species arise is referred to as speciation. The term is used most often to refer to the multiplication of species.

Many types of speciation have been proposed, but most can be grouped into three main modes. Geographic (allopatric) speciation depends upon geographic isolation of populations. Semigeographic (parapatric) speciation involves divergence between populations in continuous geographic contact. Nongeographic (sympatric) speciation involves speciation at a restricted locality, without geographic separation.

Allopatric Speciation
Geographic, or allopatric, speciation is widely accepted as the most important mode of speciation for sexual species. According to this model, a new species develops when a population becomes geographically isolated from the remainder of the species and gradually evolves independently to the extent that it becomes reproductively isolated. If the two populations reestablish contact subsequent to the development of reproductive isolation, no interbreeding will take place. Geographic speciation is a slow process not amenable to experimental testing, but abundant indirect evidence is furnished by patterns of geographic variation, the development of reproductive isolation between geographically remote elements of the same species, varying degrees of divergence between isolates, and correlation of speciation with periods of isolation produced by past climatic or geological events. An unproven variant of the allopatric model is the founder-effect model, in which an isolated population established by a few founders goes through a drastic genetic reorganization because of the limited genetic variability introduced by the founders and chance fluctuations in gene frequencies (genetic drift). Such a reorganization supposedly could accelerate the speciation process.

Parapatric Speciation
The possibility of parapatric (semigeographic) speciation is suggested by the occurrence of hybrid zones or belts along which two races interbreed freely. Many hybrid belts are undoubtedly the result of secondary contact between previously isolated populations, but others may have been formed in place. The origin of hybrid zones, interactions at hybrid zones,

and processes, if any, whereby discontinuities can proceed to reproductive isolation are subjects of continuing investigation. Most chromosomal models are also essentially parapatric. Such models usually involve fixation in a local population of chromosomal rearrangements that severely reduce fertility in heterozygotes. Small, semi-isolated populations or mating systems that promote inbreeding are usually required. The subject is controversial.

Sympatric Speciation

The origin of an asexual form from a sexual one always takes place in a nongeographic (sympatric) mode. There are many variants of asexual reproduction, including egg or seed development without fertilization (parthenogenesis and agamospermy, respectively), vegetative reproduction, and simple fission. Obligatory self-fertilization in hermaphrodites accomplishes the same end. Considerable attention has been given to the adaptive significance of uniparental reproduction and factors influencing its origin.

The only unquestioned mode of sympatric speciation for sexual forms is through polyploidy, with doubling of the entire complement of chromosomes, a phenomenon often associated with hybridization. Tetraploids (4N), for example, produced by a doubling of the diploid (2N) chromosome number, are often fertile, but are reproductively isolated from the parental forms because backcrossing produces sterile triploids (3N). Sexual polyploids are known from many animal groups but are generally rare. Polyploids are common among plants, however, and polyploidy has been an important speciation mechanism in plants.

Many other models for sympatric speciation have been proposed, and most are controversial. Many of these models involve disruptive selection, a type of selection in which two or more phenotypes have high fitness, while intermediates between them have low fitness. If an organism, for example, exploits two subniches in the same locality, and there is some impediment to free gene exchange, disruptive selection could produce two different genotypes, each adapted to one of the subniches; heterozygotes would be adapted to neither. This type of speciation has received some support from studies on herbivorous insects and insects with parasitic larvae; conditioning may favor egg deposition by the insect on the plant or host where it developed. This would inhibit free gene exchange and allow adaptation to different plants or hosts.

Ecological and Evolutionary Implications

The species is the most fundamental unit of classification and is the basic unit of reference with respect to living things. It is an important unit of interaction in natural communities. It is implicit that a species is genetically distinct from other species, with its own morphological and physiological attributes; as such it is an essential reference in all fields of biology and in the applied life sciences as well.

The origin of a new species signifies the appearance of a distinct evolutionary unit with its own potential and that is isolated by intrinsic barriers from other species.

The evolutionary implications for sexual and asexual forms, however, are quite different. Sexual forms have an immeasurably greater potential for evolutionary change because of the reservoir of genetic variability of the population, which is shared and reshuffled through sexual reproduction. Asexual forms have a limited capacity for change, rarely give rise to anything new, and tend to exploit short-term opportunities. The majority of described species other than microorganisms are sexual, although many exploit both modes of reproduction.

The speciation process itself does not necessarily produce adaptive change, although changes in adaptation are usually involved in speciation. Species, however, once formed, interact with other species through competition, and selective forces promote specialization and adaptation to subniches. This process leads not only to more diverse and complex ecological communities but also to more pronounced morphological and physiological differences between species. This phenomenon is well illustrated by various adaptive radiations that have taken place on initially depauperate islands or in species-poor lakes following colonization by one or a few species.

Speciation in this light can be seen to be a critical step that opens the way to, and indirectly promotes, evolutionary change. The origin of higher taxonomic categories has also been linked to speciation. The fossil record indicates that higher categories originate when there is movement into a major and/or distinctive and previously unexploited adaptive zone, followed by further morphological change and diversification within the zone. If a multiplicity of species are adapted to different subniches, chances are greatly enhanced for the "discovery" of a new adaptive zone by at least one species. Once movement into the zone is accomplished, diversification through speciation would take place, and interactions between species would promote further change.

John S. Mecham

See also: Adaptations and their mechanisms; Adaptive radiation; Clines, hybrid zones, and introgression; Convergence and divergence; Development and ecological strategies; Evolution: definition and theories; Evolution: history; Evolution of plants and climates; Extinctions and evolutionary explosions; Gene flow; Genetic drift; Isolating mechanisms; Natural selection; Nonrandom mating, genetic drift, and mutation; Population genetics; Punctuated equilibrium vs. gradualism; Species loss; Succession.

Sources for Further Study

Claridge, Michael F., H. A. Dawah, and M. R. Wilson, eds. *Species: The Units of Biodiversity.* New York: Chapman and Hall, 1997.

Lambert, David M., and Hamish G. Spencer, eds. *Speciation and the Recognition Concept: Theory and Application.* Baltimore: Johns Hopkins University Press, 1995.

Mayr, Ernst. *Populations, Species, and Evolution.* Cambridge, Mass.: Belknap Press, 1970.

Otte, Daniel, and John Endler, eds. *Speciation and Its Consequences.* Sunderland, Mass.: Sinauer Associates, 1989.

Paterson, H. E. H. *Evolution and the Recognition Concept of Species: Collected Writings.* Baltimore: Johns Hopkins University Press, 1993.

Townsend, Colin R., et al. *Essentials of Ecology.* 2d ed. Malden, Mass.: Blackwell Science, 2003.

White, Michael. *Modes of Speciation.* San Francisco: W. H. Freeman, 1978.

SPECIES LOSS

Types of ecology: Evolutionary ecology; Restoration and conservation ecology

Species loss, particularly the extinction of species that is caused by human activities, has increasingly concerned scientists in a number of fields, threatening biodiversity both locally and globally.

Public and scientific concern about species loss stems from several factors and encompasses a variety of viewpoints. Ethically, many people believe that species have value in and of themselves and that humankind does not have the right to cause the extinction of any species. A species may also have an unknown potential to enrich human life and health. The latter argument is important in that many synthetic medicines and commercial products were first produced by plants and animals. The loss of species could potentially mean the loss of beneficial new products for human society. Species that exist today are the result of millions of years of evolutionary success, and to lose species is to lose that evolutionary history. From a resource management point of view, ecologists and land managers alike are concerned about the effects that species loss may have on the function and stability of biotic communities.

Ecological Concerns

Relationships such as predation, competition, and parasitism link species into complex community relationships. One way species are linked is by trophic levels within the community food chain, which is more accurately described as a food web. Starting with plants at the base of the web, trophic levels begin with producers, followed by several successive levels of consumers: herbivore, first-level carnivore, second-level carnivore, and so on, up to top carnivore. Omnivores feed both as herbivores and as carnivores and thus feed at more than one trophic level. Finally decomposers feed on dead organisms and their waste products from all trophic levels.

Therefore, although the ramifications of loss of a species are not easily predicted, such a loss will have significant impact on ecosystems—often beyond merely the obvious one of increasing a population of its prey or decreasing a population of predators. Such a loss will disturb the entire food chain—or, more properly, food web—involving the complex relationships among all species in a community. Such community disturbance inevitably expands beyond the community's borders to affect the larger ecosystem.

Some examples of these complex relationships have been revealed by controlled studies.

Species-Removal Studies

Species-removal studies provide some indication of what may occur when a species becomes extinct. In more than 90 percent of predator-removal studies, population densities of prey species in the trophic level immediately below the predator have shown a significant increase or decrease. In many cases, the change in density was twofold. Rarely has the removal of predator species had no effect on the population density of its prey. However, not all studies have shown the expected increase in prey density; many have shown an unexpected decrease.

For species that possibly compete with one another, more than 90 percent of competitor-removal studies have shown an increase in the "remaining competitor" population density. Several factors may influence the strength of community response in species-removal experiments. For example, a predator may prey more heavily on a large, aggressive prey species and thus allow the coexistence of a less aggressive, competitor prey species. If the predator is removed, the aggressive prey may increase in density while the less aggressive one may actually decrease. Studies in aquatic communities indicate that the higher the trophic level in which species removal occurs, the greater the effect on population densities at lower trophic levels.

The ramifications of species loss can only partially be predicted with knowledge of community food webs. The size and direction of population density change within a community may or may not be as expected. It is safe to predict, however, that species loss will cause changes in most instances.

Wildlife Protection and Endangered-Species Legislation

Concern about species loss in North America can be traced back at least as far as 1872, when legislation offering limited protection to the American bison (buffalo) was passed by the United States Congress. This legislation was passed at the height of buffalo exploitation by market hunters and during the United States Army's policy of fighting Native American tribes by cutting off their food supply. However, President Ulysses S. Grant vetoed the legislation, and the buffalo was almost lost. Only a few hundred remained by 1900. The first National Wildlife Refuge was set aside by President Theodore Roosevelt in 1902 to protect egrets from extinction by feather hunters. Three years later, the Wichita Mountain National Wildlife Refuge was set aside to protect one of the small remnant herds of buffalo.

Species loss

Several North American species and subspecies are now extinct because of similar exploitations: The passenger pigeon, Carolina parakeet, heath hen, Merriam's elk, and Badlands bighorn sheep are some of the best known examples.

During the 1960's increasing concern about an accelerated species extinction rate attributable to human exploitation and disturbance of the environment culminated in the first federal protective legislation for endangered species, the Endangered Species Preservation Act of 1966. This act was limited to listing endangered birds and mammals and funding research on their population ecology and habitat acquisition. This legislation was expanded in 1969 to include all vertebrate animal species and some invertebrates. The definitive protection legislation is the 1973 Endangered Species Act. This act set procedures for listing threatened and endangered species, called for designation of critical habitat for each threatened or endangered species, and mandated the development of recovery plans for these species. The act prohibits the use of federal funds for projects that would harm threatened or endangered species. The coverage of the 1973 act was also expanded to include plants and invertebrate animals (except pest insects), subspecies, and distinct vertebrate populations.

Since 1966 the U.S. Fish and Wildlife Service (USFWS) has had the legal responsibility of compiling and maintaining an official threatened and endangered species list. There are formal petitioning processes for placing additional species on the list and for removing them from the list. Petitions may be initiated by the USFWS or by private organizations. Petitions are reviewed by scientific panels using all available information on the species. If sufficient information is available to support the petition, a proposed addition to the list is published in the *Federal Register* and other appropriate places to solicit public comment. Final decisions about listing, "down-listing" (for example, changing a species designation from "endangered" to "threatened"), or "de-listing" are made by the USFWS. The ultimate goal of the listing process and the implementation of a recovery plan is to increase the abundance and distribution of a species to the point of being able to remove it from the threatened and endangered species list.

James F. Fowler

See also: Biodiversity; Conservation biology; Deforestation; Endangered animal species; Endangered plant species; Extinctions and evolutionary explosions; Habitats and biomes; Old-growth forests; Reforestation; Restoration ecology; Speciation; Wildlife management; Zoos.

Sources for Further Study

Sherry, Clifford J. *Endangered Species: A Reference Handbook.* Santa Barbara, Calif.: ABC-Clio, 1998.

Stanley, Steven M. *Extinction.* New York: Scientific American Library, 1987.

Stearns, Beverly Petersen, and Stephen C. Stearns. *Watching, from the Edge of Extinction.* New Haven, Conn.: Yale University Press, 1999.

Ward, Peter D., and Don Brownlee. *Rare Earth: Why Complex Life Is Uncommon in the Universe.* New York: Copernicus, 2000.

Wilson, Edward O. *The Diversity of Life.* New York: W. W. Norton, 1993.

_____. *The Future of Life.* New York: Alfred A. Knopf, 2001.

SUCCESSION

Type of ecology: Community ecology

Succession is the progressive and orderly replacement of one biological community by another until a relatively stable, self-maintaining "climax community" is achieved.

Succession is an important ecological phenomenon because it allows the maximum variety and number of species to occupy a given area through time and leads to the establishment of an ecologically stable climax community that represents the most complex and diverse biological system possible, given existing environmental conditions and available energy input. As succession proceeds, significant changes occur in species composition, nutrient cycling, energy flow, productivity, and stratification. Changes also occur within the climax community; however, these changes act to maintain the climax, not alter it.

Immature communities tend to have high populations of a few species that are relatively small and simple. Biomass (weight of living material) is low, and nutrient conservation and retention are poor. Food chains are short, and available energy is shared by few species. Community structure is simple and easily disrupted by external forces. As communities mature, larger and more complex organisms appear, and there is a higher species diversity (number of different species). Biomass increases, and nutrients are retained and cycled within the community. The greater number of species results in more species interactions and the development of complex food webs. Community productivity (conversion of solar energy to chemical energy), initially high in immature communities, becomes balanced by community respiration as more energy is expended in maintenance activities.

Stages of Succession

The entire sequence of communities is called a sere, and each step or community in the sequence is a seral stage. The climax community is in balance, or equilibrium, with the environment and displays greater stability, more efficient nutrient and energy recycling, a greater number of species, and a more complex community structure than that of each preceding seral stage.

Each seral stage is characterized by its own distinctive forms of plant and animal life, which are adapted to a unique set of chemical, physical, and biological conditions. Excepting the climax community, change is the one constant shared by all seral stages. Changes can be induced by abiotic factors, such as erosion or deposition, and by biotic factors, modification of the environment caused by the activities of living organisms within the community.

These self-induced factors bring about environmental changes detrimental to the existing community but conducive to invasion and replacement by more suitably adapted species. For example, lichens are one of the first colonizers of barren rock outcrops. Their presence acts to trap and hold windblown and water-carried debris, thereby building up a thin soil. As soil depth increases, soil moisture and nutrient content become optimal for supporting mosses, herbs, and grasses, which replace the lichens. These species continue the process of soil-building and create an environment suitable for woody shrubs and trees.

In time, the trees overtop the shrubs and establish a young forest. These first trees are usually shade-intolerant species. Beneath them, the seeds of the shade-tolerant trees germinate and grow up, eventually replacing the shade-tolerant species. Finally, a climax forest community develops on what once was bare rock, and succession ends.

Primary and Secondary Succession
The sere just described—from barren rock to climax forest—is an example of primary succession. In primary succession, the initial seral stage, or pioneer community, begins on a substrate devoid of life or unaltered by living organisms. Succession that starts in areas where an established community has been disturbed or destroyed by natural forces or by human activities (such as floods, windstorms, fire, logging, and farming) is called secondary succession.

An example of secondary succession occurs on abandoned cropland. This is referred to as old-field succession and begins with the invasion of the abandoned field by annual herbs such as ragweed and crabgrass. These are replaced after one or two years by a mixture of biennial and perennial herbs, and by the third year the perennials dominate. Woody shrubs and trees normally replace the perennials within ten years. After another ten or twenty years have passed, a forest is established, and ultimately, after one or two additional seral stages in which one tree community replaces another, a climax forest emerges.

Both primary and secondary succession begin on sites typically low in nutrients and exposed to extremes in moisture, light intensity, temperature, and other environmental factors. Plants colonizing such sites are tolerant of harsh conditions, are characteristically low-growing and relatively small, and have short life cycles. By moderating the environmental conditions, these species make the area less favorable for themselves and more favorable for plants that are better adapted to the new environment. Such plants are normally long-lived and relatively large. Secondary succession usually proceeds at a faster rate than primary succession, because a well-developed soil and some life are already present.

Aquatic Environments
Succession can also take place in aquatic environments, such as a newly formed pond. The pioneer community consists of microscopic organisms that live in the open water. Upon death, their remains settle on the bottom and join with sediment and organic matter washed into the pond. An accumulation of sediment provides anchorage and nutrients for rooted, submerged aquatic plants such as pondweeds and waterweeds. These add to the buildup of sediment, and as water depth decreases, rooted, floating-leaved species such as water lilies prevent light from reaching the submerged aquatics and eliminate them.

At the water's edge, emergent plants rooted in the bottom and extending their stems and leaves above water (cattails, rushes, and sedges) trap

Each stage of a community's succession is characterized by its own distinctive life-forms. For example, this ancient bog in North Carolina was formed when emergent plants at the water's edge extended their stems and leaves above water, trapping sediment and adding organic matter. A soil rich in partially decomposed organic matter and saturated with water accumulated. (AP/Wide World Photos)

sediment, add organic matter, and continue the filling-in process. The shallow margins fill first, and eventually the open water disappears and a marsh or bog forms. A soil rich in partially decomposed organic matter and saturated with water accumulates. As drainage improves and the soil becomes raised above the water level, trees and shrubs tolerant of wet soils invade the marsh. These act to lower the water table and improve soil aeration. Trees suited to drier conditions move in, and once again a climax community characteristic of the surrounding area develops.

Theories of Succession

The American ecologist Frederic E. Clements (1874-1945) believed that the characteristics of a climax community were determined solely by regional climate. According to Clements, all communities within a given climatic region, despite initial differences, eventually develop into the same climax community. Some seral stages might be abbreviated or skipped entirely, while others could be lengthened or otherwise modified; however, the end result would always be a single climax community suited to the regional climate. This phenomenon is called convergence, and Clements's single-climax concept is known as the monoclimax theory.

Some ecologists have found the monoclimax theory to be simplistic and have offered other theories. One of these, the polyclimax theory, holds that, within a given climatic region, there could be many climaxes. It was noted that in any single climatic region, there were often many indefinitely maintained communities that could be considered separate and distinct climaxes. These developed as a result of differences caused by soil type, soil moisture, nutrients, slope, fire, animal activity (grazing and browsing), and other factors. Clements countered that these would eventually reach true climax status if given enough time and proposed terms such as subclimax (a long-lasting seral stage preceding the climax) and disclimax (a nonclimax maintained by continual disturbance) to describe such situations.

A third theory, the climax pattern concept, views the climax as a single large community composed of a mosaic or pattern of climax vegetation instead of many separate climaxes or subclimaxes. Numerous habitat and environmental differences account for the patterns of populations within the climax; no single factor such as climate is responsible.

While there is little doubt about the reality of succession, it is apparently not a universal phenomenon. For example, disturbed areas within tropical rain forests do not undergo a series of seral stages leading to reestablishment of the climax community. Instead, the climax is established directly by the existing species. Nevertheless, in most regions succession is the mechanism by which highly organized, self-maintained, and ecologically efficient communities are established.

Basis for Biodiversity
Succession is an important ecological phenomenon because it allows the maximum variety and number of species to occupy a given area through time and leads to the establishment of an ecologically stable climax community that represents the most complex and diverse biological system possible, given existing environmental conditions and available energy input. As succession proceeds, significant changes occur in species composition, nutrient cycling, energy flow, productivity, and stratification. Changes also occur within the climax community; however, these changes act to maintain the climax, not alter it.

Immature communities tend to have high populations of a few species that are relatively small and simple. Biomass (weight of living material) is low, and nutrient conservation and retention are poor. Food chains are short, and available energy is shared by few species. Community structure is simple and easily disrupted by external forces. As communities mature, larger and more complex organisms appear, and there is a higher species diversity (number of different species). Biomass increases, and nutrients are retained and cycled within the community. The greater number of species results in more species interactions and the development of complex food webs. Community productivity (energy storage), initially high in immature communities, becomes balanced by community respiration as more energy is expended in maintenance activities. Community structure increases in complexity and stability as equilibrium with the prevailing physical environment is achieved.

Human beings, throughout their history, have been interacting with natural communities and the process of succession—sometimes with disastrous results. Much of what was once climax forest or grassland has been put to the plow, timbered, strip-mined, or otherwise altered. In such cases, humans retard or reverse succession by destroying or disrupting the existing climax and replacing it with an ecologically simpler and less stable seral stage. In a cornfield, for example, the existing climax has been replaced, in effect, by a simple pioneer community whose dominant species is an annual. Invading weeds and shrubs must be constantly controlled or eliminated. Nutrients, in the form of fertilizer, must be applied to maintain high yields. Windstorms, drought, insect attacks, and other natural calamities can easily destroy the entire community. A cornfield is neither self-repairing nor self-maintaining, and humans must constantly intercede to keep succession in check.

Although such human intervention is easily justified by its benefits, human exploitation of the natural environment is too often destabilizing and destructive. The Dust Bowl of the 1930's is an excellent example. The shortgrass climax community that existed in parts of the southwestern Great Plains was converted to wheat production and ranchland without regard for the consequences. Drought, overgrazing, and poor farming practices combined to convert the once-fertile prairie into a barren wasteland. Failure to understand the physical and climatic conditions that resulted in shortgrass prairie, coupled with unwise land usage, was the underlying cause of this environmental disaster.

Steven D. Carey

See also: Biodiversity; Biogeography; Biomes: determinants; Biomes: types; Coevolution; Communities: ecosystem interactions; Communities: structure; Competition; Ecology: definition; Food chains and webs; Gene flow; Genetic diversity; Speciation; Symbiosis; Trophic levels and ecological niches.

Sources for Further Study

Bazzaz, F. A. *Plants in Changing Environments: Linking Physiological, Population, and Community Ecology.* New York: Cambridge University Press, 1996.

Brewer, Richard. *The Science of Ecology.* Fort Worth, Tex.: Saunders College Publishers, 1994.

Perry, David A. *Forest Ecosystems.* Baltimore: Johns Hopkins University Press, 1994.

Smith, Robert L. *Ecology and Field Biology.* 6th ed. San Francisco: Benjamin Cummings, 2001.

SUSTAINABLE DEVELOPMENT

Types of ecology: Restoration and conservation ecology; Theoretical
  ecology

*Sustainable development meets the consumption needs of the current generation
without compromising the ability of future generations to increase their economic
production to meet future needs. Environmental benefits arise as a consequence of
changes in human attitude and behavior, technology, and resource utilization.*

In 1987, the United Nations World Commission on Environment and Development, also known as the Brundtland Commission, issued a report
in which it noted that humanity has the ability to make development "sustainable"—to ensure that it meets the needs of the present without compromising the ability of future generations to meet their own needs. Sustainable development is a process of change in which the exploitation of
resources, the direction of investments, the orientation of technological development, and institutional change are all in harmony and enhance both
current and future potential to meet human needs and aspirations. The
commission envisioned the possibility of continued economic growth,
population stabilization, improvements in global economic equity between rich and poor nations, and environmental improvement, all occurring simultaneously and in harmony. Since publication of the Brundtland
Commission's report, sustainable development has become the dominant
global position on the environment, ecology, and economic development.

A Value System
Sustainable development is a normative philosophy, or value system, concerned with equal distribution of the earth's natural capital among current
and future generations of humans. Sustainable development promotes
three core values. First, current and future generations should each have
equal access to the planet's life-support systems—including Earth's gaseous atmosphere, biodiversity, stocks of exhaustible resources, and stocks
of renewable resources—and should maintain the earth's atmosphere,
land, and biodiversity for future generations. Exhaustible resources, such
as minerals and fossil fuels, are used sparingly and conserved for use by
future generations. Renewable resources, such as forests and soil fertility,
are renewed as they are used to ensure that stocks are maintained at or
above current levels and are never exhausted.

Second, all future generations should have an equal opportunity to en-

618

joy a material standard of living equivalent to that of the current generation. In addition, the descendants of the current generation in underdeveloped regions are permitted to increase their economic development to match that available to descendants of the current generation in the industrialized regions. Future development and growth in both developed and underdeveloped regions must be sustainable.

Finally, future development must no longer follow the growth path taken by the currently industrialized countries but should utilize appropriate technology. Development should also limit use of renewable resources to each resource's maximum sustained yield, the rate of harvest of natural resources such as fisheries and timber that can be maintained indefinitely through active human management of those resources.

Weak sustainability requires that depletions in natural capital be compensated for by increases in human-made capital of equal value. For example, the requirements for weak sustainability are met when a tree (natural capital) is cut for the construction of a frame house (human-made capital). However, if the tree is cut and cast aside in a land-clearing project, the requirements for weak sustainability are not met. Strong sustainability requires that depletions of one sort of natural capital be compensated for by increases in the same or similar natural capital. For example, the requirements for strong sustainability are met when a tree is cut and a new tree is planted to replace it, or when loss of acreage in equatorial rain forests in Brazil is compensated for by an increase in the acreage of temperate rain forests on the Pacific coast of North America.

Promoting the Philosophy

Sustainable development is promoted through a combination of public policies. First, to the extent possible, elements in the earth's support system are assigned monetary values in order to make the economic and financial calculations that are necessary to ensure that the requirements of weak sustainability are met.

Second, economic development in the underdeveloped world is shifted away from high-resource-using, high-polluting patterns of Western development and toward more sustainable or "appropriate" patterns. Suggested appropriate technologies include solar energy, resource recycling, cottage industry, and microenterprises (factories built on a small scale).

Third, objective and measurable air, water, and resource quality standards are established and enforced to ensure that a continuing minimum quality and quantity of natural capital is maintained and that certain stocks of natural capital are protected through the establishment of wilderness areas, oil and gas reserves, and other reserves.

Finally, each individual human adopts a personal commitment to a sustainable lifestyle, thus making a minimal personal impact on the earth's natural capital.

Environmental improvement results from the changes in resource utilization. For example, reductions in use and waste of natural capital reduces the environmental impact of resource extraction industries such as strip mines, and waste disposal industries such as incinerators. Environmental quality standards and maintenance of biodiversity leads to implementation of antipollution and ecosystem restoration efforts.

Gordon Neal Diem

See also: Biodiversity; Conservation biology; Deforestation; Endangered animal species; Endangered plant species; Erosion and erosion control; Forest management; Grazing and overgrazing; Integrated pest management; Multiple-use approach; Old-growth forests; Reforestation; Restoration ecology; Species loss; Urban and suburban wildlife; Waste management; Wildlife management; Zoos.

Sources for Further Study

Bowers, John. *Sustainability and Environmental Economics: An Alternative Text*. Harlow, England: Longman, 1997.

Dryzek, John, ed. *Debating the Earth: The Environmental Politics Reader*. New York: Oxford University Press, 1998.

Lee, Kai N. *Compass and Gyroscope: Integrating Science and Politics for the Environment*. Washington, D.C.: Island Press, 1993.

Sitarz, Daniel. *Sustainable America: America's Environment, Economy, and Society in the Twenty-first Century*. Foreword by Al Gore. Carbondale, Ill.: EarthPress, 1998.

United Nations Earth Summit. *Agenda 21*. New York: Author, 1992.

World Bank. *Monitoring Environmental Progress*. Washington, D.C.: Environmentally Sustainable Development, World Bank, 1995.

_____. *The World Bank and the Global Environment: A Progress Report*. Washington, D.C.: Author, 2000.

SYMBIOSIS

Type of ecology: Community ecology

All animals live in close association, or symbiosis, with other species. Most symbioses are based on nutritional interrelationships involving competition or cooperation. Some animals cannot survive without their symbiotic partners, while others are harmed or killed by them.

Understanding the ways in which different species of animals interact in nature is one of the fundamental goals of ecology. Predator-prey relationships, competition between species for limited resources, and symbiosis are the major forms of species interactions, and these have profoundly influenced the diversity and ecology of all forms of life. Significant advances have been made in understanding how organisms interact, but in studies of symbiosis (which literally means "living together"), one finds the most complex, interesting, and important examples of both cooperation and exploitation known in the living world.

Defining Symbiosis

Symbiosis involves many types of dependent or interdependent associations between species. In contrast to predator-prey interactions, however, symbioses are seldom rapidly fatal to either of the associating species (symbionts) and are often of long duration. With the exception of grazing animals that do not often entirely consume or destroy their plant "prey," most predators quickly kill and consume their prey. While a predator may share its prey with other individuals of the same species (clearly an example of "living together"), such intraspecific behavior is not considered to be a type of symbiosis. Fleas, some ticks, mites, mosquitoes, and other bloodsucking flies are viewed as micropredators rather than parasites.

All organisms are involved in some form of competition. The abundance and availability of environmental resources are finite, and competition for resources occurs both between members of the same species and between individuals and populations of different species. When the number of individuals in a population increases, the intensity of competition for limited food, water, shelter, space, and other resources necessary for survival and reproduction also increases. Thus, competition plays a major role in populations of free-living animals (those not inhabiting the body of other organisms) and in populations living on or in other animals. For example, both tapeworms and whales must compete for resources, and both have evolved habitat-specific adaptations to accomplish this goal. Whales compete with whales, fish, and other predators for food; tapeworms compete with tapeworms and other symbionts (such as roundworms) for food and space; and tapeworms and whales compete with each other for food in the whale's gut.

"Symbiosis" is a term used to describe nonaccidental, nonpredatory associations between species. When used by itself, the term "symbiosis" does not provide information on how or why species live together, or the biological consequences of their interactions. Recognizably different forms of symbioses all have one or more characteristics in common. All involve "living together"; most involve food sharing; many involve shelter; and some involve damage to one or both symbionts.

Hosts and Symbionts
Host species may be thought of as landlords. Hosts provide their symbionts (also called symbiotes) with transportation, shelter, protection, space, some form of nutrition, or some combination of these. Host species are generally larger and structurally more complex than their symbionts, and different parts of a host's body (skin, gills, and gut, for example) may provide habitats for several different kinds of symbionts at the same time. The three primary categories of symbiosis most commonly referred to in popular and scientific works are commensalism, mutualism, and parasitism.

Symbionts that share a common food source are known as commensals (literally, "mess-mates"). In the usual definition of commensalism, one species (usually referred to as the commensal, although both species are commensals) is said to benefit from the relationship, while the other (usually referred to as the host) neither benefits nor is harmed by the other. Adult tapeworms which live in the intestinal tracts of vertebrate hosts provide a classic example of commensals. Adult tapeworms share the host's food, usually with little or no effect on otherwise healthy hosts. As in all species, however, too large a tapeworm population may result in excessive competition, lower fitness, or disease in both the host and the tapeworms. For example, the broad fish tapeworm, which includes humans among its hosts, *Diphyllobothrium latum*, may cause a vitamin B-12 deficiency and anemia in humans when the worm burden is high. In addition to tapeworms, many human symbionts called "parasites" are, in fact, commensals.

External commensals (those living on the skin, fur, scales, or feathers of their hosts) are called epizoites. A good example of an epizoite is the fish louse (a distant relative of the copepod), which feeds on mucus of the skin and scales of fish. Another type of commensalism is called phoresis

Remoras attach themselves to larger fish for transport. This form of symbiosis is called phoresis, the passive transportation of the commensal (here, the remora) by its host. (Digital Stock)

(phoresy), which involves passive transportation of the commensal (phoront) by its host. Examples of phoreses include barnacles carried by whales and sea turtles, and remoras (sharksuckers), which, in the absence of sharks, may temporarily attach themselves to human swimmers. In inquilinism, the transported commensal (inquiline) shares, or more accurately, steals, food from the host, or may even eat parts of the host. Perhaps the best-known inquilines are the glass- or pearlfish, which take refuge in the cloacae of sea cucumbers and often eat part of the host's respiratory system. A unique type of commensalism, known as symphilism, is found in certain ants and some other insects (hosts) which "farm" aphids (symphiles) and induce them to secrete a sugary substance which the ants eat.

Mutualism

The most diverse type of commensalism is mutualism. In some works, particularly those dealing with animal behavior, mutualism is used as a synonym of symbiosis; hence, the reader must use caution in order to determine an author's usage of these terms. As used here, mutualism is a special case of commensalism, a category of symbiosis. The relationship between mutuals may be obligatory on the part of one or both species, but it is always reciprocally beneficial, as the following examples illustrate.

Some species of hermit crab place sea anemones on their shells or claws (sea anemones are carnivores which possess stinging cells in their tentacles). Hermit crabs without anemones on their shells or claws may be more vulnerable to predators than those with an anemone partner. Hermit crabs, which shred their food in processing it, lose some of the scraps to the water, which the anemones intercept, and eat. Thus, the crab provides food to the anemone, which in turn protects its provider. Such relationships, which are species-specific, are probably the result of a long period of co-evolution.

A different type of mutualism, but one having the same outcome as the crab-anemone example, is found in associations between certain clown fish and sea anemones. Clown fish appear to be fearless and vigorously attack intruders of any size (including scuba divers) that venture too close to "their" anemone. When threatened or attacked by predators, these small fish dive into an anemone's stinging tentacles, where they find relative safety. Anemones apparently share in food captured by clown fish, which have been observed to drop food on their host anemone's tentacles.

Cleaning symbiosis is another unique type of mutualism found in the marine environment. In this type of association, marine fish and shrimp of several species "advertise" their presence by bright and distinctive color patterns or by conspicuous movements. Locations where this behavior occurs are called "cleaning stations." Instead of being consumed by predatory fish, these carnivores approach the cleaner fish or shrimp, stop swimming, and sometimes assume unusual postures. Barracudas, groupers, and other predators often open their mouths and gill covers to permit the cleaners easy entrance and access to the teeth and gills. Cleaners feed on epizoites, ectoparasites, and necrotic (dead) tissue that they find on host fish, to the benefit of both species. Some studies have shown that removal of cleaning symbionts from a coral reef results in a significant decrease in the health of resident fish.

Parasitism

Parasitism is a category of symbiosis involving species associations that are very intimate and in which competitive interactions for resources may be both acute and costly. The extreme intimacy (rather than damage) between host and parasite is the chief difference between parasitism and other forms of symbiosis. Parasites often, but not always, live within the cells and tissues of their hosts, using them as a source of food. Some types of commensals also consume host tissue, but in such cases (pearl fish and sea cucumbers, for example) significant damage to the host rarely occurs. Commensalism is associated with nutritional theft.

Some, but not all, parasites harm their hosts, by tissue destruction (consumption or mechanical damage) or toxic metabolic by-products (ammonia, for example). Commonly, however, damage to the host is primarily the result of the host's own immune response to the presence of the parasite in its body, cells, or tissues. In extreme cases, parasites may directly or indirectly cause the host's death. When the host dies, its parasites usually die as well. It follows that the vast majority of host-parasite relationships are sublethal. A number of parasites are actually beneficial or crucial to the survival of their hosts. The modern, and biologically reasonable, definition of parasitism as an intimate type of symbiosis, rather than an exclusively pathogenic association between species, promotes an ecological-evolutionary understanding of interspecies associations. Most nonmedical ecologists and symbiotologists agree that two distinct forms of intimate associations, or parasitisms (with many intermediate types) occur in nature. The most familiar are those involving decreased fitness in humans and in their domestic animals and crops.

Among animal parasites, malarial parasites, hookworms, trypanosomes, and schistosomes (blood flukes) cause death and disease in millions of people each year. The degree to which these parasites are pathogenic, however, is partly the result of preexisting conditions of ill health, malnutrition, other diseases, unsanitary living conditions, overcrowding, or lack of education and prevention. Parasites which frequently kill or prevent reproduction of their hosts do not survive in an evolutionary sense, because both the parasites and their hosts perish. Both members of intimate symbiotic relationships constantly adapt to their environments, and to each other. Over time, evolutionary selection pressures result in co-adaptation (lessening of pathogenicity) or destruction or change in form of the symbiosis.

Nonpathogenic or beneficial host-parasite associations are among the most highly evolved of reciprocal interactions between species. The extreme degree of intimacy of the symbionts (not lack of pathogenicity) distinguishes this type of parasitism from mutualism. Parasitic dinoflagellates (relatives of the algae that cause "red tides") are found in the tissues of all reef-building corals. These photosynthetic organisms use carbon dioxide and other waste products produced by corals. In turn, the dinoflagellates (*Symbiodinium microadriaticum*) provide their hosts with oxygen and nutrients that the corals cannot obtain or produce by themselves. Without parasitic dinoflagellates, reef-building corals starve to death. Similar host-parasite relationships occur in termites, which, without cellulose-digesting parasitic protozoans in their gut, would starve to death.

Research on Commensals and Parasites
The life cycles of many commensals and parasites are extremely complex and often involve two or more intermediate hosts living in different environments, as well as free-living developmental stages. Knowledge of life cycles remains as one of the most important areas of research in parasitology and is usually the phase of research following the description of a new species.

Scientists have long recognized that "chemical warfare" (antibiotics, antihelminthics, insecticides) against microbial and animal parasites, and their insect and other vectors, provides only short-term solutions to the control or eradication of symbionts of medical importance. Research attempts are being made to find ways of interrupting life cycles, sometimes with the use of other parasites. This research requires sophisticated ecological and biochemical knowledge of both the host-parasite relationship and the parasite-mix. Studies of the parasite-mix are ecological (parasite-parasite and host-parasite competition), immunological (host defense mechanisms and parasite avoidance strategies), and ethological (host and symbiont behavioral interactions) in nature. Investigators involved in this kind of research must be well trained in many of the biological disciplines, including epidemiology (the distribution and demographics of disease).

Immunology is the most promising modern research area in parasitology. Not only have specific diagnostic tests for the presence of cryptic (hidden or hard to find) parasites been developed, but also vaccines may be discovered that can protect people from such destructive protozoan diseases as malaria. Malaria has killed more humans than any other disease in history, and it currently causes the death of more than one million people, and lowers the quality of life for millions of others, each year.

All species are involved in complex interrelationships with other species that live in or on their bodies, or with which they intimately interact behaviorally or ecologically. Such interactions may play a minor role in the life and well-being of one or both of the associates, or they may be necessary for the mutual survival of both. In relatively few symbiotic relationships, one or both species may suffer damage or death. Pathogenic associations are relatively rare, because disease or death of one symbiont generally results in corresponding disease or death of the other. Such relationships, which cannot persist over evolutionarily long periods of time, may nevertheless cause catastrophic loss of life in nonadapted host populations.

Ecological Implications
All species are involved in complex interrelationships with other species that live in or on their bodies, or with which they intimately interact

behaviorally or ecologically. Such interactions may play a minor role in the life and well-being of one or both of the associates, or they may be necessary for the mutual survival of both. In relatively few symbiotic relationships, one or both species may suffer damage or death. Pathogenic associations are relatively rare, because disease or death of one symbiont generally results in corresponding disease or death of the other. Such relationships, which cannot persist over evolutionarily long periods of time, may nevertheless cause catastrophic loss of life in nonadapted host populations.

Domestic animals cannot live in some parts of the world, such as the central portion of Africa, because they have little or no resistance to parasites of wild species, which are the normal hosts and are not harmed. Native species have coadapted with the parasites. This situation presents a moral dilemma to humans. In the face of human needs for space and other resources, should native animals be displaced or killed? Or should human populations proactively slow their reproductive rates? History shows that humanity has often chosen to take the former course.

The common view that animals which live in other animals are degenerate creatures that take advantage of more deserving forms of life is understandable but inaccurate. Symbionts are highly specialized animals that do not live cost-free, or always to the detriment of their hosts. Symbiotic relationships between species have vastly increased the diversity, complexity, and beauty of the living world.

Sneed B. Collard

See also: Animal-plant interactions; Biological invasions; Coevolution; Communities: ecosystem interactions; Communities: structure; Competition; Ecology: definition; Food chains and webs; Lichens; Mycorrhizae; Pollination; Predation; Trophic levels and ecological niches.

Sources for Further Study

Boothroyd, John C., and Richard Komuniecki, eds. *Molecular Approaches to Parasitology*. New York: Wiley-Liss, 1995.

Caullery, Maurice. *Parasitism and Symbiosis*. London: Sidgwick and Jackson, 1952.

Limbaugh, Conrad. "Cleaning Symbiosis." *Scientific American* 205 (August, 1961): 42-49.

Margulis, Lynn. "Symbiosis and Evolution." *Scientific American* 225 (August, 1971): 48-57.

Margulis, Lynn, and Dorion Sagan. *Slanted Truths: Essays On Gaia, Symbiosis, and Evolution*. New York: Copernicus, 1997.

Noble, Elmer, Glenn Noble, Gerhard Schad, and Austin MacGinnes. *Parasitology: The Biology of Animal Parasites*. 6th ed. Philadelphia: Lea & Febiger, 1989.

Toft, Catherine Ann, Andre Aeschlimann, and Liana Bolis, eds. *Parasite-Host Associations: Coexistence or Conflict?* New York: Oxford University Press, 1991.

Whitefield, Philip. *The Biology of Parasitism: An Introduction to the Study of Associating Organisms*. Baltimore: University Park Press, 1979.

Zann, Leon P. *Living Together in the Sea*. Neptune City, N.J.: T. F. H., 1980.

Zinsser, Hans. *Rats, Lice, and History*. Reprint. New York: Bantam Books, 2000.

TAIGA

Types of ecology: Biomes; Ecosystem ecology

"Taiga" derives from a Russian word for the forests of cone-bearing, needle-leaved, generally evergreen trees of northern Eurasia and North America. "Coniferous forest" and "boreal forest" are other names given to this biome. Some botanists include the temperate rain forests along the Pacific Coast of North America and the coniferous forests in the western mountains in the taiga.

While the term "coniferous forest" can be applied to temperate rain forest and coniferous forest biomes in the western mountains, the terms "taiga" and "boreal forest" should be restricted to the northern forests. "Taiga" is also sometimes used in a more restricted way, to mean a subdivision of the boreal forest.

Components

The dominant plants in the taiga are cone-bearing, needle-leaved, evergreen trees, such as pines, spruces, and firs. North American taiga is dominated by two species of spruce: black spruce (*Picea mariana*) and white spruce (*Picea glauca*). Jack pine (*Pinus banksiana*), balsam fir (*Abies balsamea*), and eastern larch (*Larix laricina*, a deciduous conifer) are also important in parts of the taiga. A few deciduous flowering trees are also important components. Quaking aspen (*Populus tremuloides*, the most widespread tree species in North America) and paper birch (*Betula papyrifera*) are two examples. Eurasian taiga is dominated by related species of spruce and pine and has the same character.

Determinants and Adaptations

Taiga occurs in a broad band across Canada, Alaska, Siberia, and Europe; essentially, this band is interrupted only by oceans. This pattern suggests that climate plays a major role in determining the distribution of the taiga. Average temperatures are cool, and precipitation is intermediate, but evaporation is low because of the cool temperatures. Hence, moisture is generally available to plants during the growing season. The growing season is short, and winters are long. Permafrost is present in the northern part of the taiga, and wetlands are common because drainage is often deficient. These physical conditions are primarily determined by the high latitude at which taiga occurs, but why taiga develops under these conditions is not entirely clear.

Taiga, or boreal forest, is characterized by its location in the northern latitudes and its domination by cone-bearing, evergreen trees, notably spruces and pines. A few deciduous, flowering trees are also important components, such as aspens, the most widespread tree species in the North American taiga.
(PhotoDisc)

The length of the growing season may help explain why the dominant taiga trees are evergreen. Because they retain their leaves through the winter, these trees can carry out some photosynthesis on mild winter days. More important, they avoid the energetic expense of replacing all their leaves at one time. Deciduous trees put tremendous amounts of energy into leaf replacement each spring and must replace those energy stores as well as produce energy for growth during the growing season. Deciduous forests generally occur south of the taiga, where the growing season is longer. However, some deciduous trees are successful in the taiga, so other adaptations must also be important.

Asexual reproduction probably contributes to the success of taiga trees, especially in severe environments. Black and white spruce reproduce by layering, the growth of a new tree from a lower branch which makes contact with the ground. Most deciduous trees of the taiga can sprout from the roots or other underground parts if the aboveground part of the tree is damaged or killed. Both strategies allow new trees to develop using the resources of the parent tree. In contrast, some plants growing from seed do not have sufficient resources to survive.

Fire is an important environmental factor in the taiga. Many of the conifers produce at least some cones which open and release their seeds only
after they have been heated intensely, as in a forest fire. Jack pine responds
to fire this way, as does black spruce to a lesser extent. Most deciduous
trees send up new stems from undamaged underground parts after a taiga
fire. White spruce does not employ either of these strategies but does have
efficient seed dispersal and so can move into a burned area fairly quickly.
Similar adaptations make Eurasian taiga species fit for life in northern environments. Apparently, no single suite of adaptations suits a tree species
for taiga life; instead various combinations of characteristics are employed
by the different species.

Adjacent Zones

The taiga is bordered by tundra to the north, and the meeting place between the two biomes is a broad transition zone often called the "taiga-
tundra," or forest-tundra. This ecotone is composed of a mixture of forest
and tundra plants, with trees becoming fewer and smaller from south to
north until conditions become so harsh that trees can no longer grow.

The southern boundary of the taiga is often adjacent to deciduous forest, grassland, or parkland. These are also broad, transitional ecotones. In
eastern North America, the northern hardwood forest region is such a
transition zone and is composed primarily of a mixture of trees from the
deciduous forests and the taiga. The aspen parklands in the west are also
transitional. Quaking aspen from the taiga and grasses from western grasslands mix in this zone between the taiga and grassland biomes.

Environmental Concerns

Human activities may have less impact on the taiga than on many other
biomes, primarily because the taiga occurs in a harsh environment less accessible to humans than many other biomes. Still, there are serious concerns. Acid rain became a problem for the taiga in eastern Canada in the
late twentieth century. These forests are northeast of the industrial centers
in the United States, and the prevailing southwesterly winds move nitrogen and sulfur oxides into eastern Canada, where they precipitate on
plants and soil. Both oxides interact with water to produce acids, thus acid-
ifying the soil and plant leaves. Many ecologists believe that acid precipitation has seriously damaged the taiga of both North America and Eurasia.

Global warming is a second and perhaps more insidious threat to the
taiga. The taiga will almost certainly be negatively impacted by changes in
temperature, the length of the growing season, fire frequency and intensity, and precipitation patterns. Taiga itself may play a role in carbon stor-

age and mitigation of the greenhouse effect. This possibility, its role as a source of timber, and the inherent value of the biome and its component species make it imperative that the taiga be conserved.

Carl W. Hoagstrom

See also: Biomes: determinants; Biomes: types; Chaparral; Deserts; Forests; Grasslands and prairies; Habitats and biomes; Lakes and limnology; Marine biomes; Mediterranean scrub; Mountain ecosystems; Old-growth forests; Rain forests; Rain forests and the atmosphere; Rangeland; Reefs; Savannas and deciduous tropical forests; Tundra and high-altitude biomes; Wetlands.

Sources for Further Study
Barbour, Michael G., and William Dwight Billings, eds. *North American Terrestrial Vegetation*. 2d ed. New York: Cambridge University Press, 2000.
Larsen, James A. *The Boreal Ecosystem*. New York: Academic Press, 1980.
Vankat, John L. *The Natural Vegetation of North America: An Introduction*. Malabar, Fla.: Krieger, 1992.

TERRITORIALITY AND AGGRESSION

Type of ecology: Behavioral ecology

Aggressive behavior and territoriality are common features of animals. Territories may differ in function across species, but general trends occur. Territoriality is best viewed as a means by which individuals maximize their own reproductive success rather than as a mechanism of population regulation.

Any field or forest inhabited by animals contains countless invisible lines that demarcate territories of individuals of many different species. Humans are oblivious to these boundaries yet have quick perception of human property lines; other animals are equally oblivious to human demarcations. Most organisms, in fact, appear to attend only to the territorial claims made by members of their own species. If separate maps of individual territories could be obtained for each species in the same habitat and superimposed on one another, the resulting hodgepodge of boundaries would show little consensus on the value of particular areas. Yet basic similarities exist in why and how different species are territorial.

Causes of Territoriality

The existence of aggression and territorial behavior in nature hardly comes as a surprise. Even casual observations at a backyard bird feeder reveal that species that are commonly perceived as friendly can be highly aggressive. The observation of birds at feeders can lead to interesting questions concerning territorial behavior. For example, bird feeders usually contain much more food than any one bird could eat: Why, then, are aggressive interactions so common? Moreover, individuals attack conspecifics more often than birds of other species, even when all are eating the same type of seeds.

Aggressive defense of superabundant resources is not expected to occur in nature; however, bird feeders are not a natural phenomenon. Perhaps the aggressive encounters that can be observed are merely artifacts of birds trying to forage in a crowded, novel situation, or perhaps bird feeders intensify aggressive interactions that occur less frequently and less conspicuously in nature. While the degree to which aggression observed at feeders mirrors reality is open to question, the observation of a greater intensity of interactions between conspecifics definitely reflects a natural phenomenon. Members of the same species are usually more serious competitors than are members of different species because they exploit exactly the same resources; members of different species might only share a few types of resources. Despite the ecological novelty of artificial feeders, noting which individuals win and lose in such an encounter can provide valuable information on the resource-holding potential of individuals that differ in various physical attributes such as body size, bill size, or even sex. For organisms that live in dense or remote habitats, this type of information can often be obtained only by observations at artificial feeding stations.

Territorial defense can be accomplished by visual and vocal displays, chemical signals, or physical encounters. The sequence of behaviors that an individual uses is usually predictable. The first line of defense may involve vocal advertisement of territory ownership. One function of birdsong is to inform potential rivals that certain areas in the habitat are taken. If song threats do not deter competitors, visual displays may be employed. If visual displays are also ineffective, then residents may chase intruders and, if necessary, attack them. This sequence of behaviors is common in territorial interactions because vocal and visual displays are energetically cheaper than fighting and involve less risk of injury to the territory owner.

It may be less obvious why fighting is a necessary component in territorial interactions for both territory owners and intruders. Without the threat of bodily injury, there is no cost to intruders that steal the resources of an-

Two elephants exhibit territorial aggression. (Digital Stock)

other individual. This would severely hamper an owner's ability to control an area. On the other hand, if intruders never physically challenge territory owners, then it would pay for all territory owners to exaggerate their ability to defend a resource. Thus, physical aggression may be essential. Animals do not frequently kill their opponents, however, so there must be something that limits violence. Various species of animals possess formidable weapons, such as large canine teeth or antlers, that are quite capable of inflicting mortal wounds. Furthermore, a dead opponent will never challenge again. Yet fights to the death are rare in nature. When they do occur, some novel circumstance is usually involved, such as a barrier that prohibits escape of the losing individual. Restraint in normal use of weapons, however, probably does not indicate compassion among combatants. Fights to the death may simply be too costly, because they would increase the chance that a victor would suffer some injury from a loser's last desperate attempts to survive.

Functions of Territory

Territories can serve various functions, depending on the species. For some, the area defended is only a site where males display for mates; for others, it is a place where parents build a nest and raise their offspring; for others, it may be an all-purpose area where an owner can have exclusive access to food, nesting sites, shelter from the elements, and refuge from predators. These different territorial functions affect the area's size and the length of time an area is defended. Territories used as display sites may be only a few meters across, even for large mammal species. Territorial nest sites may be smaller still, such as the densely packed nest sites guarded by parents of many colonial seabirds. All-purpose territories are typically large relative to the body size of the organism. For example, some passerine birds defend areas that may be several hundred meters across. Although all three types of territories may be as ephemeral as the breeding season, it is not uncommon for all-purpose territories to be defended year around.

The abundance and spatial distribution of needed resources determine the economic feasibility of territoriality. On one extreme, if all required resources are present in excess throughout the habitat, territory holders should not have a reproductive advantage over nonterritory holders. At the other extreme, if critical resources are so rare that enormous areas would have to be defended, territory holders might again have no reproductive advantage over nonterritory holders. If needed resources, however, are neither superabundant nor extremely rare and are somewhat clumped in the habitat, territoriality might pay off. That is, territorial individuals might produce more offspring than nonterritorial individuals.

Studies of territoriality raise more questions than biologists can answer. Researchers investigate how large an area an individual defends and whether both sexes are equally territorial. They seek to determine whether the territories of different individuals vary according to quality. The density of conspecifics may influence territoriality; on the other hand, territoriality itself may serve to regulate population size, although evidence suggests that this is an incidental effect.

All-purpose territories vary considerably in size, depending on the resource requirements of the individuals involved and the pattern of temporal variation in resource abundance. In some organisms, individuals only defend enough area to supply their "minimum daily requirements." In others, individuals defend a somewhat larger area—one that could still support them even when resource levels drop. In others, individuals defend territories that vary in size depending on current resource levels. For example, pied wagtails (European songbirds) defend linear territories along riverbanks that are about six hundred meters long during the winter. The emerging aquatic insects they consume are a renewable resource, but renewal rates vary considerably during the season. Rather than adjusting territory size to match the current levels of prey abundance in the habitat, wagtails maintain constant territory boundaries. This inflexibility persists even though territories that extend for only three hundred meters could adequately support an individual for about one-third of the season. In contrast, the territory size of an Australian honey eater varies widely during the winter. Nectar productivity of the flowers visited by honey eaters varies considerably during the season. By adjusting territory size to match changing resource levels, individual birds obtain a relatively constant amount of energy each day (about eighteen kilocalories).

Sex Roles

In some species, only males are territorial. In other species, both sexes defend territories, but males defend larger territories than females do. In some mammals in which both sexes are territorial, males are aggressive only to other males, and females are only aggressive to other females. In these species, male territories are sufficiently large to encompass the territories of several females. Presumably, these males have increased sexual access to the females within their territories. Perhaps the most curious example of sex-specific territorial behavior is observed in a number of coral reef fish, in which all individuals in the population are initially female and not territorial. As the individuals grow older and larger, some develop into males. Once male, they engage in territorial behavior.

Within a species, significant variation in territory quality exists among individuals. Studies on numerous species have demonstrated a relationship between territory quality and an individual's resource-holding potential. For example, larger individuals tend to control prime locations more often than smaller individuals. In addition, possession of higher-quality territories often results in increased reproductive success. For some species, this occurs because individuals with better territories obtain mates sooner or obtain more mates than individuals with poorer territories. In other species, possession of superior territories increases the survival chances of the owner.

As the density of conspecifics increases, the ability of individuals to control territories decreases. In some species, the territorial system may break down completely, with all individuals scrambling for their share of needed resources in a chaotic fashion. In other species, the territorial system is replaced by a dominance hierarchy. All competitors may remain in the area, but their access to resources is determined by their rank in the hierarchy. For example, dominant male elephant seals can successfully defend from other males areas containing between eighty and one hundred females. Very dense clusters of females, however (two hundred or more), attract too many males for one male to monopolize. When this happens, one male—usually the largest male—dominates the rest and maintains disproportionate access to females.

Territoriality undeniably has an adaptive function: to increase the survival and reproductive success of individuals. Territoriality can also have several possible incidental effects, one of which was once considered to be an adaptive function: serving as a means of population regulation. The reasoning behind this hypothesis is simple. The number of territories in a habitat would limit the number of reproducing individuals in a population and would thereby prevent overpopulation that could cause a population crash. Support for this hypothesis would include demonstration that a significant number of nonbreeding adults exist in a population. Indeed, for several species, experimental removal of territory owners has revealed that "surplus" individuals quickly fill the artificially created vacancies. In most of the species studied, however, these surplus individuals are primarily males. Population growth can be curbed only by limiting the number of breeding females, not the number of breeding males. Furthermore, the population regulation argument assumes that some individuals abstain from reproduction for the good of the population. If such a population did exist, a mutant individual that never abstained from reproducing would quickly spread, and its descendants would predominate in future generations.

Territoriality in the Field

Territoriality is typically investigated in the field using an observational approach. Initial information collected includes assessing the amount of area used by each individual, how much of that area is defended from conspecifics, and exactly what is being defended. It is relatively easy to discern the spatial utilization of animals. For many species, all that is required is capturing each individual, marking it for field identification, and watching its movements. For species that range long distances, such as hawks or large mammals, and species that are nocturnal, radio telemetry is frequently used. This methodology requires putting radio transmitters on the individuals to be followed and using hand-held antennas, or antennas attached to cars or airplanes, to monitor movements. For fossorial species (animals that are adapted for digging), animal movements are often determined by repeated trapping. This method involves placing numerous baited live traps above the ground in a predetermined grid.

Knowing the spatial utilization of an animal does not document territoriality. Many types of animals repeatedly use the same regions in the habitat but do not defend these areas from conspecifics. Such "home ranges" may or may not contain areas that are defended (that is, territories). Territorial defense can be readily documented for some animals by simply observing individual interactions. These data often need to be supplemented by experiments. Behavioral interactions might occur only in part of the organism's living space because neighbors do not surround it. For these individuals, researchers play tape-recorded territorial vocalizations or place taxidermy mounts of conspecifics in different locations and note the response of the territory holder. For other species, such as fossorial rodents, direct estimates of territory size cannot be obtained because aggressive interactions cannot be observed; as a result, territory boundaries must be inferred from trapping information. Regions in which only the same individual is repeatedly trapped are likely to be areas that the individual defends. This is an indirect method, however, and can be likened to watching the shadow of an organism and guessing what it is doing.

It is often difficult to determine exactly what an animal is defending in an all-purpose territory where organisms use many different types of resources. Which resource, that is, constitutes the "reason" for territorial defense? On the other hand, several resources may contribute in some complex way. For many species these things simply are not known. This uncertainty also complicates estimates of territory quality. For example, red-winged blackbirds in North America have been particularly well studied for several decades by different investigators in various parts of the species range. Males defend areas in marshes (or sometimes fields), and some males obtain significantly more mates than others. Biologists think that males defend resources that are crucial for female reproduction. Some males may be more successful at mating than others because of variation in territory quality. Yet the large number of studies done on this species has not yielded a consensus on what the important resources are, whether food, nest sites, or something else.

Theoretical investigations of territorial behavior often employ optimality theory and game theory approaches. Optimality theory considers the benefits and costs of territorial defense for an individual. Benefits and costs might be measured simply as the number of calories gained and lost, respectively. Alternatively, benefits might be measured as the number of young produced during any one season; costs might be measured as the reduction in number of future young attributable to current energy expenditures and risks of injury. For territorial behavior to evolve by means of natural selection, the benefits of territorial behavior to the individual must exceed its costs.

Game theory analyses compare the relative success of individuals using alternative behaviors (or "strategies"). For example, two opposing strategies might be "defend resources from intruders" and "steal resources as they are encountered." In the simplest case, if some individuals only defend and other individuals only steal resources, the question would be which type of individual would leave the most offspring. Yet defenders interact with other defenders as well as with thieves, and the converse holds for thieves. By considering the results of interactions within and between these two types of individuals, a game theory analysis can predict the conditions under which one strategy would "win" or "lose" and how the success of each type of individual would vary as the frequency of the other increases in the population. A complete understanding of territoriality involves not only empirical approaches in the field but also the development of testable theoretical models. Considerable advances have been made recently merging these two methodologies. Future investigations will no doubt include experimental control over resource levels that will allow definitive tests of predictions of alternative theoretical models.

Evolution of Terrioriality
Among animals in general, some species are highly aggressive in defending their living space, and others ignore or tolerate conspecifics in a nearly utopian manner. Some animals are territorial during only part of the annual cycle, and some only in specific areas that they inhabit; others remain aggressive at any time and in any place. Thus, a main goal for researchers is

to unravel the ecological and evolutionary conditions that favor aggressive behavior and territoriality.

Aggression and territorial behavior appear to have evolved in various organisms because, in the past, aggressive and territorial individuals outreproduced nonaggressive and nonterritorial ones.

An implicit assumption of behavioral biologists is that animals other than humans do not interact aggressively because of conscious reasoning, nor are they consciously aware of the long-term consequences of aggressive acts. Should these consequences be detrimental, natural selection will eliminate the individuals involved, even if this means total extinction of the species. Humans are different. They are consciously aware of their actions and of the consequences of such actions. They need only use conscious reasoning and biological knowledge of aggressive behavior to create conditions that can reduce conflict between individuals and groups.

Richard D. Howard

See also: Altruism; Communication; Defense mechanisms; Displays; Ethology; Habituation and sensitization; Hierarchies; Insect societies; Mammalian social systems; Mimicry; Pheromones; Predation; Reproductive strategies.

Sources for Further Study

Alcock, John. *Animal Behavior.* 7th ed. Sunderland, Mass.: Sinauer Associates, 2001.

Allen, Colin, and Marc Bekoff. *Species of Mind: The Philosophy and Biology of Cognitive Ethology.* Cambridge, Mass.: MIT Press, 1997.

Davies, Nicholas B., and John R. Krebs. *An Introduction to Behavioral Ecology.* 4th ed. Boston, Mass.: Blackwell Scientific Publications, 1997.

Dennen, J. van der, and V. S. E. Falger, eds. *Sociobiology and Conflict: Evolutionary Perspectives on Competition, Cooperation, Violence, and Warfare.* New York: Chapman and Hall, 1990.

Howard, Eliot. *Territory in Bird Life.* New York: Atheneum, 1962.

Ratcliffe, Derek A. *The Peregrine Falcon.* 2d ed. San Diego, Calif.: Academic Press, 1993.

Wilson, Edward O. *Sociobiology.* Cambridge, Mass.: The Belknap Press of Harvard University Press, 1975.

TROPHIC LEVELS AND ECOLOGICAL NICHES

Types of ecology: Community ecology; Ecoenergetics; Ecosystem ecology

A trophic level is a position in the food pyramid occupied by an organism based on its food relationships with other organisms: what it eats, and what eats it. An ecological niche is the physical space in which an animal lives and all the interactions with the other living organisms and components of its environment.

The idea of the niche probably had its first roots in ecology in 1910. At that time, Roswell Johnson wrote that different species utilize different niches in the environment. He theorized that individuals of a particular species are only in certain places because of food supply and environmental factors that limit their distribution in an area. Later, in 1924, Joseph Grinnel developed his concept of niche that centered on an organism's distribution having limits set on it by climatic and physical barriers. At the same time, Charles Elton was defining his own idea of niche. His description of niche involved the way an organism makes its living—in particular, how it gathers food.

The Food Pyramid

For many years, ecologists focused on Elton's definition and referred to niche in terms of an organism's place in the food pyramid. The food pyramid is a simplified scheme in which organisms interact with one another while obtaining food. The food pyramid is represented as a triangle, often with four horizontal divisions, each division being a different trophic level.

The base of the food pyramid is the first trophic level and contains the primary producers: photosynthetic plants. At the second trophic level are the primary consumers. These are the herbivores, such as deer and rabbits, which feed directly on the primary producers. Secondary consumers are found at the third trophic level. This third trophic level contains carnivores, such as the mountain lion. The members of the uppermost trophic level are the scavengers and decomposers, including hyenas, buzzards, fungi, and bacteria. The organisms in this trophic level break down all the nutrients (such as carbon and nitrogen) in the bodies of plants and animals and return them to the soil to be absorbed and used by plants.

It should be noted that no ecosystem actually has a simple and well-defined food pyramid. Many organisms interact with more than only the organisms at the adjacent trophic levels. For example, a coyote could be considered to belong to the third trophic level with the carnivores, but the coyote also feeds on occasional fruits and other primary producers. Basically, all living things are dependent on the first trophic level, because it alone has the capability to convert solar energy to energy found in, for example, glucose and starch. The food pyramid takes the geometric form of a triangle to show the flow of energy through a system.

Photosynthetic plants lose 10 percent of the energy they absorb from the sun as they convert solar energy into glucose and starch. In turn, the herbivores can convert and use only 90 percent of the energy they obtain by eating plants. Hence, less energy is found at each higher trophic level. Because of this reduced energy, fewer organisms can be supported by each higher trophic level. Consequently, the sections of the pyramid get smaller at each higher trophic level, representing the decreasing levels of energy and number of members.

Types of Niches

Through the years, two concepts of niche have evolved in ecology. The first is the place niche, the physical space in which an organism lives. The second is the ecological niche, and it encompasses the particular location occupied by an organism and its functional role in the community.

The functional role of a species is not limited to its placement along a food pyramid; it also includes the interactions of a species with other organisms while obtaining food. For example, the methods used to tolerate the physical factors of its environment, such as climate, water, nutrients, soils, and parasites, are all part of its functional role. In other words, the ecological niche of an organism is its natural history: all the interactions and interrelationships of the species with other organisms and the environment.

The study of the interrelationships among organisms has been the focus of ecological studies since the 1960's. Before this time, researchers had focused on the food pyramid and its effect on population changes of merely a single species. One example, the classic population study of the lynx and the snowshoe hare of Canada, originally focused on the interactions of the species in the food pyramid. It was discovered that the lynx had a ten-year population cycle closely following the population cycle of its prey, the snowshoe hare. The lynx population appeared to rise, causing a decline in the population of the snowshoe hare. In the investigations that followed, however, studies diverted the focus from the food pyramid to other ele-

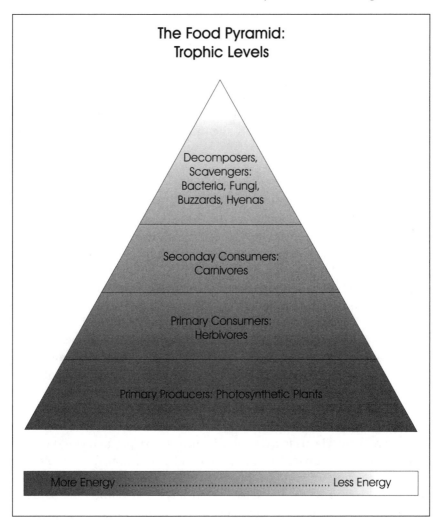

The Food Pyramid:
Trophic Levels

Decomposers,
Scavengers:
Bacteria, Fungi,
Buzzards, Hyenas

Seconday Consumers:
Carnivores

Primary Consumers:
Herbivores

Primary Producers: Photosynthetic Plants

More Energy .... Less Energy

ments of the niche of the two species. For example, the reproductive nature of the hare provided a contradiction to the simple predator-prey explanation. The hare has a faster rate of reproduction than the lynx. It seemed impossible that the significantly lower population of lynx could effectively place sufficient predator pressure on the hare to cause its drastic decline in numbers. Therefore, it appeared that the population dynamics of the hare and lynx were regulated by more than simply a predator-prey relationship.

Later studies of the lynx and hare suggested that the peaks and dives in the two populations may also be a factor of parasites of the hare that are carried by the lynx. A rise in the lynx population increases the carriers of parasites of the hare. Therefore, it is thought that, although the hare has a much greater reproduction rate than the lynx, the population of hares will still decline because of the combination of predation by the lynx and the increased frequency of parasites of the hare. This study involved looking at more than one dimension of the ecological niche of a species and broke away from concentrating on only the interactions between organisms in the food pyramid.

Niche Overlap: Interspecific
The goal of understanding how species interact with one another can also be better accomplished by defining the degree of niche overlap, the degree of the sharing of resources between two species. When two species use one or more of the same elements of an ecological niche, they exhibit inter-specific competition. It was once believed that interspecific competition would always lead to survival of only the better competitor of the two species. That was the original concept of the principle of the competition exclusion law of ecology: No two species can utilize the same ecological niche. It was conjectured that the weaker competitor would either migrate, begin using another resource not used by the stronger competitor, or become extinct. It is now believed that the end result of two species sharing elements of ecological niches may not always be exclusion.

Ecologists theorize that similar species do, in fact, coexist, despite the sharing of elements of their ecological niches, because of character displacement, which leads to a decrease in niche overlap. Character displacement involves a change in the morphological, behavioral, or physiological state of a species without geographical isolation. Character displacement occurs as a result of natural selection arising from competition between one or more ecologically similar species. Examples might be changes in mouth sizes so that they begin to feed on different sizes of the same food type, thereby decreasing competition.

Specialists and Generalists
The more specialized a species, the more rigid it will be in terms of its ecological niche. A species that is general in terms of its ecological niche needs will be better able to find and use an alternative for the common element of the niche. Since a highly specialized species cannot substitute whatever is being used, it cannot compete as well as the other species. Therefore, a specialized species is more likely to become extinct.

For example, a panda is a very specialized feeder, eating mainly bamboo. If a pest is introduced into the environment that destroys bamboo, the panda will probably starve, being unable to switch to another food source. On the other hand, the coyote is a generalized feeder. A broad variety of food types make up its diet. If humans initiate a pest-control program, killing the population of rabbits, the coyote will not fall victim to starvation, because it can switch to feeding predominantly on rodents, insects, fruits, and domesticated animals (including cats, dogs, and chickens). Hence, species with specialized ecological niche demands (specialists) are more in danger of extinction than those with generalized needs (generalists). Although this fundamental difference in survival can be seen between specialists and generalists, it must be noted again that exclusion is not an inevitable result of competition. There are many cases of ecologically similar species that coexist.

Niche Overlap: Intraspecific
When individuals of the same species compete for the same elements of the ecological niche, it is referred to as intraspecific competition. Intraspecific competition has the opposite effect of interspecific competition: niche generalizations. In increasing populations, the first inhabitants will have access to optimal resources. The opportunity for optimal resources decreases as the population increases; hence, intraspecific competition increases. Deviant individuals using marginal resources may slowly begin to use less optimal resources that are in less demand. That can lead to an increase in the diversity of ecological niches used by the species as a whole. In other words, the species may become more generalized and exploit wider varieties of niche elements.

Representing a situation on the opposite end of the spectrum from that of two organisms competing for the same dimension of an ecological niche is the vacant niche theory. This ecological principle states that when an organism is removed from its ecological niche, space, or any other dimension of the niche, another organism of the same or similar species will reinvade.

Field Research
Theoretical studies of ecological niches are abstract, since humans are limited to three-dimensional diagrams, and there are more dimensions than three to an ecological niche. This multidimensionality is referred to as the *n*-dimensional niche. This abstract n-dimensional niche can be studied mathematically and statistically, but the study of ecological niches is mainly a field science. Therefore, its techniques are mainly those used for field research.

Research that attempts to describe all the elements of the n-dimensional ecological niche would require extensive observations. Yet, ecological niches are difficult to measure not only because of the plethora of data that would have to be collected but also because of the element of change in nature. The internal and external environment of an organism is always dynamic. Nothing in life is static, even if equilibrium is established.

These constant fluctuations create daily and seasonal changes in space and ecological niches. Therefore, because of the constant fluctuations, merely descriptive field observations would not be reliable depictions of an organism's ecological niche. Ecologists must also resort to quantitative data of measurable features of an organism's ecological niche. For example, the temperature, pH, light intensity, algae makeup, predators, and activity level of the organism are measurable features of an ecological niche in a pond community. The difficulty is in the collection of each of the necessary measurements making up an ecological niche. The ecologist would have to limit the data to a manageable number of specific dimensions of the niche based on conjecture and basic intuition. Such limitations often lead to incomplete and disconnected measurements that can at best only partially describe a few of the dimensions of the ecological niche.

Ecologists realize that complete observations and measurements of all the dimensions of an organism's ecological niche are unattainable. The focus in understanding how a species interacts with its community centers on determining the degree of niche overlap between any two species. In other words, the level of competition for space niche and resources. Studies of this niche overlap are typically limited to dimensions that can be quantitatively measured. Yet, there is still the problem of deciding which of the dimensions are involved in the competition between the two species. Again, the ecologist must usually rely on inherent knowledge about the two species in question. Often, researchers investigating niche competition measure no more than four ecological niche dimensions to determine the niche overlap in an attempt to understand how two individuals competing for the same space, resources, or other ecological niche features can coexist.

Field methods for observations and quantitative measurements of elements of ecological niches, niche overlap, and niche competition are probably endless. To name a few, describing an organism's niche may involve fecal samples to determine its diet, fecal samples of possible predators to identify its primary predator, animal and plant species checklists of its space niche along with soil components, climatic trends, and the like. Niche competition and overlap often can be studied first in the laboratory under controlled situations. One method might involve recording the population dynamics of the species as different elements in the ecological niche are manipulated to determine which is the better competitor and what is the resource that is most responsible for limiting the population size.

Niche and Community

The shift in meaning and study from merely space and trophic level placement in the food pyramid to ecological niche of n dimensions has been beneficial for the field of ecology. This focus on community ecology is obviously much more productive for the goal of ecology, the understanding of how all living organisms interact with one another and with nonliving elements in the environment.

Perhaps more important is the attempt to describe niches in terms of community ecology, which can be essential for some of humankind's confrontations with nature. For example, it has become increasingly apparent that synthetic chemicals are often too costly and too hazardous to continue using for control of crop pests and carriers of diseases. The goal is to control pests effectively with biological controls. Biological controls can involve the introduction of natural predators of the undesirable pest or the introduction of a virus or bacteria that eliminates the pest and is harmless to humans and wildlife.

The success of a biological control is directly proportional to the knowledge of the pest's n-dimensional ecological niche and the other organisms with which it comes in contact. A classic example of the havoc that can result from manipulations of nature without adequate ecological information is when Hawaii attempted to use biological controls to eradicate a population of snakes, which humans had accidentally introduced. The biological control used was the snake's natural predator, the mongoose. One very important dimension of the ecological niche of both species was ignored. One species was active only at night, while the other was active only during the day. Needless to say, this particular venture with a biological control was not a success.

Another relevant function of community-oriented studies of ecological niches involves endangered species. In addition to having aesthetic and potential medicinal values, an endangered organism may be a keystone species, a species on which the entire community depends. A keystone species is so integral to keeping a community healthy and functioning that if obliterated the community no longer operates properly and is not productive.

Habitat destruction has become the most common cause of drastic population declines of endangered species. To enhance the habitat of the endangered species, it is undeniably beneficial to know what attracts a species to its particular preferred habitat. This knowledge involves the details of many of the dimensions of its ecological niche integral to its population distribution. Another common means of endangering the survival of a species is to introduce an organism or exotic species that competes for the same resources and displaces the native species. Solving such competition between native and introduced species would first involve determining niche overlap.

It is often stated that an ounce of prevention is worth a pound of cure. Thus, the researching and understanding of all the dimensions of ecological niches are key to preventing environmental manipulations by humankind that might lead to species extinction. Many science authorities have agreed that future research in ecology and related fields should focus on solving three main problems: species endangerment, soil erosion, and solid waste management.

This focus on research in ecology often means that studies of pristine communities, those undisturbed, will be the most helpful for future restoration projects. Although quantitative and qualitative descriptions of pristine areas seem to be unscientific at the time they are made, because there is no control or experimental group, they are often the most helpful for later investigations. For example, after a species has shown a drastic decline in its population, the information from the observations of the once-pristine area may help to uncover what niche dimension was altered, causing the significant population decrease.

Jessica O. Ellison

See also: Animal-plant interactions; Balance of nature; Biodiversity; Biogeography; Biological invasions; Coevolution; Communities: ecosystem interactions; Communities: structure; Competition; Food chains and webs; Herbivores; Omnivores; Phytoplankton; Predation; Symbiosis.

Sources for Further Study

Bronmark, Christopher, and Lars-Anders Hansson. *The Biology of Lakes and Ponds.* New York: Oxford University Press, 1998.

Ehrlich, Paul R. "Who Lives Together, and How." In *The Machinery of Nature.* New York: Simon & Schuster, 1986.

Giller, Paul S. *Community Structure and the Niche.* New York: Chapman and Hall, 1984.

Odling-Smee, F. John, Kevin N. Laland, and Marcus W. Feldman. "Niche Construction." *American Naturalist* 147, no. 4 (April, 1996): 641-649.

Rayner, Alan D. M. *Degrees of Freedom: Living in Dynamic Boundaries.* River Edge, N.J.: World Scientific Publications, 1997.

Ricklefs, Robert E. *Ecology*. 4th ed. New York: Chiron Press, 1999.

Shugart, Herman H. *Terrestrial Ecosystems in Changing Environments*. New York: Cambridge University Press, 1998.

Smith, Robert L. *Ecology and Field Biology*. 6th ed. San Francisco: Benjamin/Cummings, 2001.

Stone, Richard. "Taking a New Look at Life Through a Functional Lens." *Science* 269, no. 5222 (July, 1995): 316-318.

TROPISMS

Type of ecology: Physiological ecology

Tropisms represent a variety of adaptations of plants that have allowed them to grow toward or away from environmental stimuli such as light, gravity, objects to climb, moisture in soil, or the position of the sun. These rapid responses, which make use of separate systems for detecting and responding to stimuli, help a plant to survive in its particular habitat.

Although plants appear not to move, they have evolved adaptations to allow movement in response to various environmental stimuli; such mechanisms are called tropisms. There are several kinds of tropism, each of which is named for the stimulus that causes the response. For example, gravitropism is a growth response to gravity, and phototropism is a growth response to unidirectional light. Tropisms are caused by differential growth meaning that one side of the responding organ grows faster than the other side of the organ. This differential growth curves the organ toward or away from the stimulus. Growth of an organ toward an environmental stimulus is called a positive tropism; for example, stems growing toward light are positively phototropic. Conversely, curvature of an organ away from a stimulus is called a negative tropism. Roots, which usually grow away from light, are negatively phototropic. Tropisms begin within thirty minutes after a plant is exposed to the stimulus and are usually completed within approximately five hours.

Phototropism

Phototropism is a growth response of plants to light coming from one direction. Positive phototropism of stems results from cells on the shaded side of a stem growing faster than cells along the illuminated side; as a result, the stem curves toward the light. The rapid elongation of cells along the shaded side of a stem is controlled by a plant hormone called auxin that is synthesized at the stem's apex. Unidirectional light causes the auxin to move to the shaded side of stems. The increased amount of auxin on the shaded side of stems causes cells there to elongate more rapidly than cells on the lighted side of the stem. This, in turn, causes curvature toward the light.

Only blue light having a wavelength of less than 500 nanometers can induce phototropism. The photoreceptors in this system are called cryprochromes and may alter the transport of auxin across cellular membranes, thereby facilitating its transport to the shaded side of the stem. Phototropism is important for two main reasons: It increases the probability of stems and leaves intercepting light for photosynthesis and of roots obtaining water and dissolved minerals that they need.

Gravitropism

Gravitropism is a growth response to gravity. The positive gravitropism of roots involves the root cap, a tiny, thimble-shaped organ approximately 0.5 millimeter long that covers the tip of roots. Decapped roots grow but do not respond to gravity, indicating that the root cap is necessary for root gravitropism. Gravity-perceiving cells, called columella cells, are located in the center of the root cap. Each columella cell contains fifteen to twenty-five amyloplasts (starch-filled plastids) which, under the influence of gravity, sediment to the lower side of columella cells. This gravity-dependent sedimentation of amyloplasts is the means whereby roots sense gravity, possibly by generating electrical currents across the root tip. These gravity-induced changes are then transmitted to the root's elongating zone, lo-

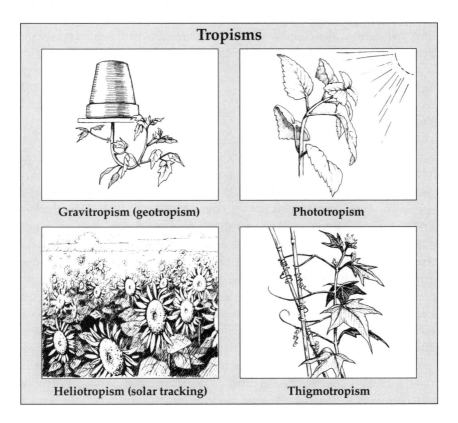

Tropisms

Gravitropism (geotropism)

Phototropism

Heliotropism (solar tracking)

Thigmotropism

cated 3 to 6 millimeters behind the root cap. The differential growth that causes curvature occurs in the elongating zone.

When roots are oriented horizontally, growth along the lower side of the elongating zone is inhibited, thereby causing the root to curve downward. Among the first events that produce this differential growth is the accumulation of calcium ions along the lower side of the root tip. Calcium ions move to the lower side of the cap and elongating zone of horizontally oriented roots. This movement may be aided by electrical currents in the root. The accumulation of calcium along the lower side of the root causes the auxin to accumulate there as well. Because auxin inhibits cellular elongation in roots, the lower side of the root grows slower than the upper side of the root, and the root curves downward. When the root becomes vertical, the lateral asymmetries of calcium and auxin disappear, and the root grows straight down.

Gravity-sensing cells in stems are located throughout the length of the stem. As in roots, the auxin and calcium ions in stem cells direct the negative gravitropism (in this case, upward curvature) of shoots. As auxin accumulates along the lower side, calcium ions gather along the upper side of horizontally oriented stems. The accumulation of auxin along the stem's lower side stimulates cellular elongation there. Gravitropism increases the probability of two important results: Roots will be more likely to encounter water and minerals, and stems and leaves will be better able to intercept light for photosynthesis.

Thigmotropism

Thigmotropism is a growth response of plants to touch. The most common example of thigmotropism is the coiling exhibited by specialized organs called tendrils. Tendrils are common on twining plants such as morning glory and bindweed. Prior to touching an object, tendrils often grow in a spiral. This type of growth is called circumnutation, and it increases the tendril's chances of touching an object to which it can cling. Contact with an object is perceived by specialized epidermal cells on the tendril. When the tendril touches an object, these epidermal cells control the differential growth of the tendril. This differential growth can result in the tendril completely circling the object within five to ten minutes. Thigmotropism is often long-lasting. For example, stroking one side of a tendril of garden pea for only a few minutes can induce a curling response that lasts for several days. Thigmotropism is probably controlled by auxins and ethylene, as these regulate thigmotropic-like curvature of tendrils even in the absence of touch.

Growing tendrils touched in the dark do not respond until they are illuminated. This light-induced expression of thigmotropism may indicate a requirement for adenosine triphosphate (ATP), as ATP will substitute for light in inducing thigmotropism of dark-stimulated tendrils. Tendrils can store the sensory information received in the dark, but light is required for the coiling growth response to occur. Thigmotropism by tendrils allows plants to "climb" objects and thereby increases their chances of intercepting light for photosynthesis.

Hydrotropism and Heliotropism

Roots also grow toward wet areas of soil. Growth of roots toward soil moisture is called hydrotropism. Roots whose caps have been removed do not grow toward wet soil, suggesting that the root cap is the site of moisture perception by roots. Hydrotropism is probably controlled by interactions of calcium ions and hormones such as the auxins.

Heliotropism, or "solar tracking," is the process by which plants' organs track the relative position of the sun across the sky, much like a radio telescope tracks stars or satellites. Different plants have different types of heliotropism. The "compass" plants (*Lactuca serriola* and *Silphium laciniatum*) that grow in deserts orient their leaves parallel to the sun's rays, thereby decreasing leaf temperature and minimizing desiccation. Plants that grow in wetter regions often orient their leaves perpendicular to the sun's rays, thereby increasing the amount of light intercepted by the leaf. Heliotropism occurs in many plants, including cotton, alfalfa, and beans. Sunflowers get their name from the fact that their flowers follow the sun across the sky. On cloudy days, leaves of many heliotropic plants become oriented horizontally in a resting position. If the sun appears from behind the clouds late in the day, leaves rapidly reorient themselves—they can move up to 60 degrees in an hour, which is four times more rapid than the movement of the sun across the sky. Heliotropism is controlled by many factors, including auxins.

Practical Implications

Biologists are studying tropisms in hopes of being able to mimic these detection and "guidance" systems for human use. Scientists at the National Aeronautics and Space Administration (NASA), for example, have been studying the way plants perceive and respond to gravity to learn how to grow plants in deep space. Understanding the gravity detection and guidance systems in plants may help people design more effective rockets, which, like plants, must respond to gravity to be effective.

Randy Moore

See also: Adaptations and their mechanisms.

Sources for Further Study

Campbell, Neil A., and Jane B. Reece. *Biology.* 6th ed. San Francisco: Benjamin Cummings, 2002.
Evans, Michael L., Randy Moore, and Karl H. Hasenstein. "How Roots Respond to Gravity." *Scientific American* 255 (December, 1986): 112-119.
Hart, James Watnell. *Plant Tropisms and Other Growth Movements.* Boston: Unwin Hyman, 1990.
Haupt, W., and M. E. Feinleib, eds. *The Physiology of Movements.* New York: Springer-Verlag, 1979.
Salisbury, Frank B., and Cleon W. Ross. *Plant Physiology.* 4th ed. Belmont, Calif.: Wadsworth, 1992.
Satter, R. L., and A. W. Galston. "Mechanisms of Control of Leaf Movements." *Annual Review of Plant Physiology* 32 (1981): 83-103.
Taiz, Lincoln, and Eduardo Zeiger. *Plant Physiology.* 2d ed. Sunderland, Mass.: Sinauer Associates, 1998.

TUNDRA AND HIGH-ALTITUDE BIOMES

Types of ecology: Biomes; Ecosystem ecology

Regions where no trees grow because of frozen soil or extreme water runoff due to steep grades (at high altitudes) are known as tundra. High altitude biomes have similar limitations on the growth of plant life.

Tundra landscapes appear where long, cold winters, a permanently frozen subsoil, and strong winds combine to prevent the development of trees. The resulting landscapes tend to be vast plains with low-growing forbs and stunted shrubs. Vast areas of this biome encircle the northernmost portions of North America and Eurasia, constituting the Arctic tundra. Climatic conditions atop high mountains at all latitudes are similar; these small, isolated areas are called the alpine tundra.

Permafrost

The low temperatures of the tundra regions cause the formation of a permanently frozen layer of soil known as permafrost. Characteristic of Arctic tundra, permafrost, which varies in depth according to latitude, thaws at the surface during the brief summers. As the permafrost below is impenetrable by both water and plant roots, it is a major factor in determining the basic nature of tundra.

The alternate freezing and thawing of soil above the permafrost creates a symmetrical patterning of the land surface characteristic of Arctic tundra. Perhaps the best known features of the landscape are stone polygons that result when frost pushes larger rocks toward the periphery, with smaller ones occupying the center of each unit. This alteration of the tundra landscape, called cryoplanation, is the major force in molding Arctic tundra landscapes.

In contrast, alpine tundra generally has little or no permafrost. Even though alpine precipitation is almost always higher than for Arctic tundra, steep grades result in a rapid runoff of water. Alpine soils are, therefore, much drier, except in the flat alpine meadows and bogs, where conditions are more like those of Arctic areas.

Vegetation

Both Arctic and alpine tundra regions are composed of plants that have adapted to the same generally stressful conditions. Biodiversity of both plants and animals—the total number of species present—is low compared to most other ecosystems. Plant growth is slow because of the short growing seasons and the influence of permafrost. Most tundra plants are low-growing perennials that reproduce vegetatively rather than by seed. Often they grow in the crevices of rocks that both shelter them in the winter and reflect heat onto them in summer.

Common plants of the low-lying Arctic tundra sites include various sedges, especially cottongrass, and sphagnum moss. On better-drained sites, biodiversity is higher, and various mosses, lichens, sedges, rush species, and herbs grow among dwarfed heath shrubs and willow. The arrangement of plants within a small area reflects the numerous microclimates resulting from the peculiar surface features.

Alpine plants possess many of the features of Arctic plants. However, because strong winds are such a prominent feature of the alpine environment, most of the plants grow flat on the ground, forming mats or cushions.

Below alpine tundra and south of Arctic tundra, there is the boreal (also known as taiga) biome, dominated by coniferous forest. Between the forest and tundra lies a transitional zone, or ecotone. This ecotone is characterized by trees existing at their northern (or upper) limit. Especially in alpine regions, stunted, gnarled trees occupy an area called krummholz. In North

A caribou herd moving across a tundra landscape in the Opingivik area of Baffin Island, Canada, in 1999. (AP/Wide World Photos)

America, the krummholz is much more prominent in the Appalachian Mountains of New England than in the western mountains.

Conservation

Like all world biomes, tundra regions are subject to degradation and destruction, especially as a result of human activities. Because of low human population density and their unsuitability for agriculture, tundras generally are less impacted by humans than are grasslands and forests. However, tundra ecosystems, when disturbed, recover slowly, if at all. As most tundra plants lack the ability to invade and colonize bare ground, the process of ecological succession that follows disturbances may take centuries. Even tire tracks left by vehicles can endure for decades. The melting of permafrost also has long-lasting effects.

The discovery of oil and gas in tundra regions, such as those of Alaska and Siberia, has greatly increased the potential for disturbances. Heavy equipment used to prospect for fossil fuels and to build roads and pipelines has caused great destruction of tundra ecosystems. As the grasses and mosses are removed, the permafrost beneath melts, resulting in soil erosion. The disposal of sewage, solid wastes, and toxic chemicals poses special problems, as such pollutants tend to persist in the tundra environment longer than in warmer areas.

Animals of the Arctic tundra, such as caribou, have been hunted by the native Inuit using traditional methods for centuries without an impact on populations. The introduction of such modern inventions as snowmobiles and rifles has caused a sharp decline in caribou numbers in some areas.

Although efforts at restoring other ecosystems, especially grasslands, have been quite successful, tundra restoration poses difficult problems. Seeding of disturbed Arctic tundra sites with native grasses is only marginally successful, even with the use of fertilizers. In alpine tundra, restoration efforts have been somewhat more successful but involve transplanting as well as seeding and fertilizing. A recognition of natural successional patterns and long-term monitoring is a necessity in such efforts.

Thomas E. Hemmerly

See also: Biomes: determinants; Biomes: types; Forests; Habitats and biomes; Mountain ecosystems; Taiga.

Sources for Further Study

Johnson, Rebecca L. *A Walk in the Tundra*. Minneapolis: Carolrhoda Books, 2000.

Sayre, April Pulley. *Tundra*. Frederick, Md.: Twenty-first Century Books, 1995.

Shepherd, Donna W. *Tundra*. New York: Franklin Watts, 1997.

Smith, R. L., and T. M. Smith. *Elements of Ecology*. 4th ed. San Francisco: Benjamin Cummings, 1998.

Walker, Tom. *Caribou: Wanderer of the Tundra*. Portland, Oreg.: Graphic Arts Center, 2000.

Zwinger, Ann H., and Beatrice E. Willard. *Land Above the Trees: A Guide to American Alpine Tundra*. Boulder, Colo.: Johnson Books, 1996.

URBAN AND SUBURBAN WILDLIFE

Types of ecology: Landscape ecology; Restoration and conservation
ecology

The global increase in the human population growth and density has seen a simul-
taneous increase in the growth of urban and suburban areas throughout the world.
As urban habitats and their suburban extensions have become more common, the
wildlife of these human landscapes has become the focus of attention and study.

Animals and plants of cities and suburbs are categorized as urban
wildlife. As cities and suburbs grow ever larger and displace natural
habitats, many city and suburban landscapes have become more attractive
for certain kinds of wildlife, or at least urban wildlife has become more noticeable. Urban wildlife consists of an eclectic and unlikely mix of escaped
pets (including cats, dogs, reptiles, exotics, and caged birds), feral animals,
furtive and temporary intruders from adjacent natural habitats, and species whose natural ecology and behavior enable them to fit within human-
modified landscapes and tolerate living in close proximity to humans.

Urban Habitats

Urban landscapes present a seemingly stark and forbidding environment
for wildlife. The horizontal streets and sidewalks are punctuated by rising
angles and arches of concrete and steel that in turn are topped by wood and
metal rooftops. Overhead, a maze of telephone, power, and cable lines limits vertical movement, while vehicle and foot traffic poses a constant threat
to surface movement. All of these edifices and connecting corridors and
lines result in a complex, vertically structured environment that some animals find difficult to maneuver yet to which other animals quickly adapt.

In addition to this monotonous and often dangerous structural diversity, urban wildlife is subject to elevated and often almost continuous noise
and disturbance and is constantly exposed to an enormous variety of residential wastes (garbage, litter, excess water, salts, sewage), vehicular pollutants (lubricants, greases, gasoline, hydrocarbons, nitrogen oxides), and
chemical wastes (pesticides, paints, lead, mercury, contaminants).

Despite the forbidding features of urban habitats, a surprising variety of
wildlife manages to exist on a more or less permanent basis. In fact, some
kinds of wildlife can be found even in the midst of the most degraded forms
of urban blight. *Ailanthus*, which is also commonly called tree-of-heaven, is
but one of many opportunistic trees and shrubs that can take root and

grow given a bare minimum of soil and nutrients. A simple linear crack in the pavement of a sidewalk, a little-used roadway, an unused parking area, or a vacant lot can trap enough windswept dirt to offer a growing substrate for *Ailanthus* and similar hardy plants. Each *Ailanthus*, in turn, provides food and shelter for equally tough and adaptable wildlife, ranging from the variety of invertebrates that colonize and feed upon *Ailanthus* to birds and mammals that take shelter or find food in its branches and foliage. Similarly, every invading sprig of grass, wildflower, shrub, or tree, however large or small, creates its own suite of microhabitats which, in turn, offer colonization opportunities for other plants and animals, the whole ultimately contributing to an overall increase in urban biodiversity.

Benefits of Urban Habitats

Ailanthus is an example of those plants and animals able to tolerate the most extreme urban conditions, but in reality most urban wildlife derives a number of benefits by living within the confines of cities and suburbs. Far from being homogeneous expanses of concrete, most urban centers are a patchwork of different habitats—residential, commercial, and industrial buildings, warehouses, power stations, vacant lots, detached gardens, rooftop gardens, and alleyways—that each offer innumerable opportunities for wildlife. Many urban areas also have a number of limited access areas that animals are quick to adopt for shelter and breeding places; these include fenced-in lots and boarded-up buildings, along with a rabbit warren of underground tunnels, ducts, steam and water pipes, basements, and access ways.

City lights extend foraging time and opportunities, allowing wildlife to hunt for food not only throughout the day but also during much of the night, as needed. Urban nooks and crannies offer an extensive variety of microhabitats that differ fundamentally in size, microclimate, and structural features. These microhabitats serve primarily as shelters and breeding sites for city wildlife. Many birds, such as house sparrows (*Passer domesticus*) and Eurasian starlings (*Sturnis vulgaris*) nest in innumerable crevices, cracks, nooks, niches, and sheltered rooftops. Pigeons (*Columbia livia*) and starlings hide in sheltered enclaves offered by bridge abutments and supports, archways, and other edifices.

The most adaptable forms of wildlife are quick to find and take advantage of subtle advantages offered in urban habitats: Many birds cluster around chimneys and roof reflectors or in shelters afforded by lee sides of rooftops during harsh cold and windstorms. Others are equally quick to obtain warmth by sitting on poles, rooftops, or other elevated perches to orient toward sunlight, while at ground level animals gather near gratings, vents, and underground heating pipes.

Urban Scavengers

Urban wildlife just as quickly concentrates in areas where potential food is made available—for instance, during trash pickup—then just as quickly disperses to find new food sources. Most forms of urban wildlife forage opportunistically as scavengers, specializing in finding and consuming all bits of discarded food, raiding trash cans, and concentrating at waste collection and disposal centers. Thus, the rubbish dumps, found in or immediately adjacent to every city of the world, attract an amazing diversity of small mammals and birds. Feeding on the scavenged food of urban areas and bird feeders is much more efficient because it requires less energy to find or catch and is usually available throughout the year.

Because of the need to find and exploit temporary food resources, some of the most successful urban animals forage in loose groupings or flocks: The more eyes there are for searching, the more feeding opportunities can be identified and exploited. Solitary and nonsocial species often do less well in urban environments simply because they lack the collective power of the group to find food and shelter, and avoid enemies.

The availability of a year-round food supply—however tenuous and temporary—along with the presence of an enormous variety of safe shelters and breeding sites promotes a higher life expectancy, which partly or mostly balances the higher vehicle-related death rates to which urban wildlife is continuously subject.

Urban Parks

Parks and open space provide the only true natural habitat refuges set deep within urban and suburban landscapes. Such open-space habitats function as ecological islands in a sea of urbanism. Most are necessarily managed habitats rather than entirely natural and, like the urban environment that surrounds them, are usually subject to constant disturbance from adjacent traffic, noise, and other forms of pollution. Economically, since most open-space parks are set aside and maintained for a variety of recreational purposes rather than as natural habitats, the wildlife that colonizes these unnatural natural habitats must have an unusually high tolerance for human presence and recreational activities of all kinds.

Urban Birds

For some forms of urban wildlife, the urban landscape is merely a human-made version of their natural environment. Thus, for pigeons the ledges, cracks, and crevices of buildings and bridges represent an urban version of the cracks and crevices of cliffs and rock outcrops that they use for roosting and nesting in their native habitats. Similarly, the short-eared owls (*Asio flammeus*) and snowy owls (*Nyctea scandiaca*) that show up in winter to stand as silent sentinels at airports, golf courses, and other open areas are simply substituting these managed short-grass habitats for the tundra habitats preferred by snowy owls and the coastal marshes hunted by short-eared owls. Their summer replacements include a host of grassland nesting species such as grasshopper sparrows, kildeer, and upland sandpipers, which all find these managed habitats to be ideal substitutes for the native grasslands which they displaced or replaced.

Many bird inhabitants of urban and suburban environments are exotics which were deliberately or inadvertently introduced into urban areas. Certainly the three birds with the widest urban distribution in North America, the pigeon or rock dove, European starling, and house sparrow or English sparrow, all fit within this category. The introduction of the European starling into North American cities and suburbs resulted from the dedicated efforts of the American Acclimitization Society of the late 1800's. The goal of this society was the successful introduction of all birds mentioned in the works of Shakespeare into North America. Unfortunately for North Americans, the character of Hotspur in *Henry IV* makes brief note of the starling, so the society repeatedly attempted to introduce the starling into Central Park until they were finally successful. Since then, the starling has become the scourge of cities and suburbs throughout much of North America and the rest of the civilized world. Starlings damage and despoil crops, and dirty buildings with their droppings.

The association between house sparrows and urban centers is apparently very old. Evidence suggests that they abandoned their migratory ways to become permanent occupants of some of the earliest settlements along the Nile and Fertile Crescent, a trend that has continued to this day. Sparrows and starlings both share certain characteristics that enable them effectively to exploit urban and suburban habitats; both are aggressive colonizers and competitors, able to feed opportunistically on grains, crops, discarded bits of garbage, and other food supplies.

Avian occupants also include an increasing diversity of released caged pets, avian and otherwise. Thus, urban locales in Florida, Southern California, and along the Gulf Coast support an ever increasing diversity of parakeets, parrots, finches, and lovebirds, all stemming from caged pet birds either deliberately released or lost as escapees.

Feral Animals

Feral animals, mostly dogs (*Canidae*) and cats (*Felidae*), represent another important source and component of urban wildlife. Feral dogs revert to

Along with squirrels, opossums, rats, coyotes, deer, bears, and other animals, the omnivorous raccoons are among the most familiar forms of wildlife that coexist with humans in urban and suburban neighborhoods, eating everything they can find and often so accustomed to the human presence that they approach people to beg for handouts. (PhotoDisc)

primal adaptive behaviors, gathering in loose packs that usually forage and take shelter together, but have limited success because almost all cities in developed countries have ongoing measures to control and remove them whenever found. Feral cats are often more successful because they are secretive, mostly nocturnal, and can better exploit available urban food sources. The role of other feral animals as urban wildlife, mostly escaped pets, is not well known.

Humans and Urban Wildlife
The attitude of urban dwellers toward urban wildlife varies greatly. For many humans, urban wildlife offers a welcome respite from their otherwise dreary and mundane surroundings. Urban wildlife in all of its forms and colors can be aesthetically attractive, even beautiful, and is also compellingly interesting. For example, the nesting of a pair of red-tailed hawks (*Buteo jamaicensis*) in New York City's Central Park sparked a remarkable interest in bird-watching in the city and a heightened awareness of exactly how exciting wildlife watching can be. All facets of the pair's courtship and nesting were observed and reported in newsprint, novellas, and even

a book, *Red-Tails in Love*. Other animals, while not nearly as large, conspicuous, and glamorous in their color and disposition, also elicit interest. Urban wildlife adds lively color and contrast to the otherwise monotonous gray and grime of streets and sidewalks. Part of the attraction is that urban birds are usually already sufficiently tolerant to be semitame in spirit, easily seen and observed, and, in some instances, easily attracted by strategically placed bird feeders and birdhouses.

Public attitudes toward urban predators vary considerably. Some people find them attractive and interesting and even put out food for them. Others consider them pests or potentially dangerous and avoid them. During rabies outbreaks or public scares, most urban wildlife is targeted by various control programs to remove unwanted animals.

Suburban Wildlife Habitats

The vast sprawl of suburbs across the landscape offers many types of wildlife yet another habitat to exploit, either as residences or as waystations during the search for food or shelter. Like urban areas, suburbs offer a range of differing habitats. The simplest suburbs are merely extensions of urban row houses with minimal yardscapes, but there is an increasing progression toward more open and natural yards in outlying suburbs that merge with rural areas and natural habitats. The larger and more diverse yards at the edges of suburbs often help blur the distinction and diversity between human landscapes and natural landscapes.

Ornamental trees, shrubs, flowers, gardens, and lawns that characterize almost all suburban habitats provide a series of artificial habitats that can actually increase wildlife diversity. Again, the chief wildlife benefactors are species that can best ecologically exploit the unnatural blend of woodland, edge, and meadow that suburban landscapes offer. It is no accident that some of the most common components of suburban wildlife include thrushes such as robins, finches, cardinals, titmice, blue jays, crows, and many other similar birds. All of these species are actually responding to the structural components of the suburban landscape, which provide suitable substitutes for their natural landscapes.

The blend of ornamental and garden vegetation offered by most suburban landscapes offers food for a diversity of what were once considered less tolerant wildlife. Deer, wild turkey, grouse, and a host of other animals, large and small, make periodic forays into suburbs in search of food. Crepuscular and nocturnal wildlife is much more likely effectively to exploit food sources offered by suburban landscapes than diurnal wildlife, which is more at risk because of its high visibility during daylight hours.

Predators

Well-wooded suburban habitats that attract a variety of wildlife also attract an increasing number of predators. American kestrels (*Falco sparverius*), Cooper's hawks (*Accipiter cooperi*), barn owls (*Tyto alba*), screech owls (*Otus* spp.), and little owls (*Athene noctua*) provide but a small sampling of birds of prey that nest deep within urban and suburban environments, taking advantage of open-space habitats deep within cities and quickly exploiting unused areas within most suburbs. Terrestrial predators are almost equally common, but most are nocturnal or nearly so; consequently, their contacts with humans are quite limited. Many urban predators, are, in fact, mistaken for neighborhood pets and left alone or are recognized and avoided: Coyotes (*Canis latrans*) are often mistaken for dogs, especially when seen in twilight. The wily coyote is equally at home in the suburbs of Los Angeles, California, and the urban parks of New Haven, Connecticut, joining a host of small and medium-sized mammal predators such as foxes (*Vulpes* spp.) and scavengers such as opossums (*Didelphis marsupalis*), raccoons (*Procyon lotor*), and skunks (*Mephitis mephitis*). These urban predators have many behavioral attributes in common. All are omnivorous and able to feed on a wide variety of natural foods such as fruit, small birds and mammals, insects, and invertebrates such as beetles, grasshoppers, and earthworms.

Foraging and food habits of urban predators sometimes conflict with human concerns. Urban foxes hunt and kill cats, especially kittens, if given the opportunity, while the larger and stronger urban coyote will often not hesitate to kill and eat cats and dogs, to pet owners' dismay.

Wildlife Management

Urban wildlife must be much more closely managed than wildlife of natural environments because urban and suburban habitats attract an enormous number of pest species as well as interesting and beneficial species. Introduced species such as starlings may also transmit histoplasmosis, a fungal disease that attacks human lungs. Other birds may also be harbingers, carriers, and vectors of various diseases, the most notable of which are the parrots and parakeets, which transmit parrot fever or psittocosis. Rats and mice (*Rodentia*) carry and spread disease and despoil both residential and public buildings and other structures.

The growing interest in urban wildlife has stimulated innumerable programs to promote beneficial wildlife. Both public and private organizations and agencies have embarked on a variety of programs aimed at remodeling existing habitats and even creating new habitats for urban wildlife.

Programs aimed at creating new or modifying existing urban habitats come in a variety of categories, such as linear parks, greenways, urban wildlife acres programs, backyard gardens, and treescaping streets and roadways, all of which create biodiversity, which in turn provides attractive habitats for colonization by additional animals and plants. Modifications of existing habitats to increase animal biodiversity include "critter crossings," roadside habitats, backyard gardens, arbor plantings, all of which provide refuges, shelters, breeding sites, connecting corridors, and safe havens that promote the welfare of urban and suburban wildlife.

Many existing open-space habitats are also being modified. Many urban renewal commissions have placed new and more restrictive regulations on the use of pesticides and fertilizers on golf courses, which not only reduces incidence and intensity of nonpoint pollution from the golf courses but also reduces the incidence of wildlife poisoning. These steps cannot help but increase the biotic potential of golf courses for supporting local biodiversity.

Dwight G. Smith

See also: Biodiversity; Food chains and webs; Landscape ecology; Trophic levels and ecological niches; Wildlife management.

Sources for Further Study
Adams, Lowell W. *Urban Wildlife Habitats: A Landscape Perspective*. Minneapolis: University of Minnesota Press, 1994.
Bird, David, Daniel Varland, and Juan Josè Negro, eds. *Raptors in Human Landscapes*. San Diego, Calif.: Academic Press, 1996.
Forman, Richard, and Michel Godron. *Landscape Ecology*. New York: John Wiley & Sons, 1986.
Gill, Don, and Penelope Bonnett. *Nature in the Urban Landscape: A Study of City Ecosystems*. Baltimore: York Press, 1973.
McDonnell, Mark J., and Steward T. A. Pickett, eds. *Humans as Components of Ecosystems*. New York: Springer-Verlag, 1993.

WASTE MANAGEMENT

Types of ecology: Ecotoxicology; Restoration and conservation ecology

Waste management concerns the physical by-products of human activity that cannot be reintegrated into the ecological biomass cycle. These by-products include solid, liquid, and airborne substances that are potentially harmful to living organisms. As the human population grows and the use of manufactured materials expands, disposing of waste becomes more challenging.

According to the World Watch Institute, world production of manufactured materials (not counting recycled materials) increased nearly 2.5 times between the early 1960's and the late 1990's. In industrialized countries the increase is far greater: The United States, for example, has seen an eighteenfold increase in materials production since 1900. The average U.S. citizen throws away an estimated 2 to 8 pounds of garbage daily, and although studies demonstrate that the per-person production of waste remained approximately the same throughout the twentieth century, the sharp rise in population and expanding industrial base meant greater total accumulations of waste. Furthermore, the types of waste changed.

Waste is commonly categorized as domestic, or solid, waste, and industrial, or liquid, waste, although the distinction is not absolute. Both may contain toxic substances, but the percentage of toxins in industrial waste is likely to be higher, and the types of waste are disposed in different ways. The smoke emitted from industrial processing of materials and vehicle exhaust are additional types of waste, although they are commonly thought of as pollution rather than waste.

Solid Waste

Solid waste is the familiar garbage that households and businesses in the United States have sent to the dump since garbage collection began late in the nineteenth century. The largest portion, more than 40 percent, consists of paper products, especially newspaper and containers. Yard waste, food debris, plastic containers and wrappings, bottles, metals, and appliances are also regularly thrown away. About 1 percent of this waste involves hazardous materials, typically insecticides, beauty aids, and cleaning products. Construction waste accounts for a large share—about 12 percent—of solid waste and may contribute a higher proportion of hazardous materials, such as solvents and paint.

Although most of these materials are solid, when dumped together

A wastewater treatment plant collects the domestic and industrial effluent that has been conveyed to its location by a sewage system and treats this contaminated water by removing solids, filters, biological decomposition, and other processes, ending in release into the ground or, more usually, into a surface watercourse. (PhotoDisc)

they can soak up rainwater and then ooze chemical-laden liquids. This leachate may filter down into the groundwater and pollute nearby streams and wells. If it contains toxic elements, such as the lead or mercury from batteries, the leachate can be dangerous to health. The odor from rotting garbage may also foul the air, seldom enough to be harmful but still repellent to people living nearby. It can attract animal scavengers, which may become infected with diseases from the garbage and spread them to other animals or even humans, especially if feces are part of the waste.

In order to combat these effects, sanitary landfills place a plastic lining under the waste to contain leachate and cover each day's load of garbage under a thin layer of soil. Pipe systems also disperse methane gas produced by rotting organic materials. The landfills are therefore less dangerous to human health or the environment, but many old, abandoned sites were not so well engineered. They may continue to dribble harmful chemicals into groundwater for decades and emit methane, which is flammable. Numerous small, illegal dumps and litter compound the problem.

Measures to reduce the amount of waste deposited in landfills have partially succeeded. Recycling has drastically cut the total paper, metal, and glass waste in some U.S. states and industrialized countries. The use of garbage disposals and composting has caused the proportion of organic materials to decline. However, such reductions did not eliminate solid waste. By the end of the twentieth century, cities were finding it increasingly difficult to find room for new landfill sites, even when the space was urgently needed. Stringent regulations about the geological composition of landfills reduced the number of usable sites, while objections from citizen action committees, known as "not in my back yard" (NIMBY) groups, also eliminated sites near populated areas.

Facilities used to incinerate waste, which sometimes powered electrical generators with the resulting heat energy, also faced objections because burning could release health-threatening materials, such as dioxins, into the air. Moreover, a significant proportion of waste, such as appliances and concrete, cannot be eliminated by burning. Tires, too hazardous to burn, float to the surface in landfills, causing continuous problems for waste managers; they often end up stacking the tires in immense piles that, if accidentally ignited, can burn out of control and create large clouds of black fumes.

Industrial Waste
The effluent stream of by-products from factories, as well as chemical and petroleum refineries, is made up of water, solid filings and cuttings, liquid solvents and oil derivatives, and semisolid sludge. The solid components are usually no more hazardous than household wastes, although medical waste—particularly tainted blood and used "sharps," such as needles and scalpels—may pose the additional danger of spreading disease. However, liquids and semiliquids sometimes contain a high proportion of hazardous chemicals. Rain also leaches chemicals, such as cyanide and mercury, out of the smelted tailings from mines. Agricultural fertilizers and pesticides can enter groundwater or streams as well. Because these liquid wastes rapidly spread through waterways and groundwater, they are often collectively known as toxic waste.

Industry now uses all ninety-two naturally occurring elements on the periodic table, and the isotopes of some of these are radioactive. Nuclear weapons manufacturing in particular leaves radioactive debris, but medical procedures that use radioactive tracers and scientific instruments may also create radioactive wastes. This nuclear waste continues to emit radiation for thousands or hundreds of thousands of years, and improperly stored radioactive materials have been associated with increased risk of disease for people, animals, and plants.

During the 1980's and 1990's federal and state regulations brought industrial waste management under rigorous control. Facilities known as secure landfills are designed to contain nonradioactive industrial wastes in tightly lined, self-contained areas. Incinerators reduce the waste to harmless ash while releasing few or no harmful particulates into the atmosphere. Separate repositories store nuclear wastes deep underground in leak-proof containers.

The public is seldom reassured by such measures, however. Leakage occasionally occurs from secure landfills. Near-zero toxic emissions from incineration means that some toxins do, in fact, escape into the atmosphere. In addition, nuclear repositories may not be catastrophe proof; for example, an earthquake could crack open containers, releasing radioactive material into groundwater supplies. Although waste managers insist that these dangers are minimal, the news media bring them to public attention, and NIMBYs regularly resist the opening of new secure landfills and radioactive waste repositories. State governments often object as well, as was the case when the Nevada legislature stalled the construction of a nuclear repository at Yucca Mountain. Many old facilities, built before strict government oversight, remain in use and could leak toxic materials into the environment undetected. The memory of deadly chemical leaks, such as that discovered at Love Canal in New York in 1976, and of released radioactive material, such as the plutonium that escaped the Hanford Nuclear Reservation in Washington State, makes the public wary of hazardous wastes.

As a result of citizen concern, most new hazardous waste disposal sites are now located far from population centers. This has created a new peril. The waste must be transported, primarily by trucks and trains, to a facility. Traffic accidents and train derailings en route can dump extremely dangerous chemicals straight into water or the atmosphere. Evacuations of residents near such accidents, while not common, increased during the 1990's. Even if people are rescued, however, plant and animal life is not safeguarded.

Environmental Consequences
Many critics of waste management insist that only source reduction—a drastic decrease in the use of raw materials—will make waste disposal safe. Accordingly, during the 1990's some countries, notably Denmark and Germany, sought to reduce virgin material use as much as 90 percent by intensifying recycling. In the United States, Superfund legislation was passed to set aside federal funds to pay for cleanups of the most dangerous hazardous waste sites. Other industrial countries have similar projects. Still, only a fraction of sites receive attention, and until source reduction goals are met, household and industrial wastes will continue to swell land-

fills with environmentally hazardous substances. Illegal dumping of hazardous waste exacerbates the danger.

Scientists disagree about how severely wastes damage the environment, but there is agreement that repercussions are evident and likely to increase. Methane from dumps, smoke-stack emissions, and vehicle exhaust contain greenhouse gases, which are implicated in global warming. Nutrients released from sewers, as well as runoff from agriculture and mining, degrade the environment of rivers and streams, harming aquatic life and leaving the water unusable without special treatment. The waterborne wastes that reach the ocean, supplemented by ocean dumping of toxic materials, alter and sometimes destroy offshore ecosystems, as is the case for many coral reefs worldwide.

Roger Smith

See also: Biological invasions; Biomagnification; Biopesticides; Deforestation; Eutrophication; Genetically modified foods; Integrated pest management; Invasive plants; Ocean pollution and oil spills; Ozone depletion and ozone holes; Pesticides; Phytoplankton; Pollution effects; Slash-and-burn agriculture.

Sources for Further Study
Baarschers, William H. *Eco-Facts and Eco-Fiction: Understanding the Environmental Debate*. New York: Routledge, 1996.
Dunne, Thomas, and Luna B. Leopold. *Water in Environmental Planning*. San Francisco: W. H. Freeman, 1978.
Gourlay, K. A. *World of Waste: Dilemmas of Industrial Development*. New York: St. Martin's Press, 1992.
Laak, Rein. *Wastewater Engineering Design for Unsewered Areas*. Lancaster, Pa.: Technomic, 1986.
McGhee, Terrence. *Water Supply and Sewerage*. New York: McGraw-Hill, 1991.
Qasim, Syed R. *Wastewater Treatment Plants: Planning, Design, and Operation*. 2d ed. Lancaster, Pa.: Technomic, 1999.
Rathje, William, and Cullen Murphy. *Rubbish! The Archaeology of Garbage*. Tucson: University of Arizona Press, 2001.
Salvato, Joseph A. *Environmental Engineering and Sanitation*. 4th ed. New York: Wiley, 1992.
Tillman, Glenn M. *Wastewater Operations: Troubleshooting and Problem Solving*. Chelsea, Mich.: Ann Arbor Press, 1996.
Whitaker, Jennifer Seymour. *Salvaging the Land of Plenty: Garbage and the American Dream*. New York: William Morrow, 1994.

WETLANDS

Types of ecology: Biomes; Ecosystem ecology

Wetlands, transitional areas between aquatic and terrestrial habitats, are home to a variety of flood-tolerant and salt-tolerant plant species.

Wetlands represent one of the most biologically unique and productive of all natural habitats. In their unaltered state, these water-influenced areas are used by a variety of wildlife species. These habitats also have the ability to take up and store water during floods, and their soils and plants have the ability to remove nutrients and heavy metals from water. The recognition of these values helped to slow the rate of wetlands loss to such uses as agricultural development and urban expansion. A desire to protect remaining wetland acres has led to a significant movement for wetlands preservation.

Definition of Wetlands

Ecologists recognize wetlands as a type of ecotone. Ecotones are unique areas that represent a transition from one type of habitat to another. Often, these transitional areas have characteristics of both habitats. Wetlands are areas located between aquatic, water-based habitats and dry land. Because they are located at the edge of an aquatic habitat, wetlands are always influenced to some degree by water. They are not always under water, as are aquatic habitats, and they are not always dry, as are terrestrial habitats.

The most important environmental factor in wetlands is the periodic or frequent occurrence of water. This presence of water influences both the nature of the soil and the flora and fauna of a region. Soils which experience periodic coverage with water become anoxic, develop a dark color, and give off an odor of hydrogen sulfide. These soil characteristics differ from those of upland soils and give wetland soils their unique hydric nature. In these soils influenced by water, only flood-tolerant hydrophyte species can exist. Hydrophytic plants vary in their tolerance to flooding from frequent (such as bald cypress) to infrequent (such as willows).

In defining a particular area as a wetland, often all three of the components listed above are used: water, hydric soils, and hydrophytic plants. However, the presence of water is not always a reliable indicator because water rarely covers a wetland at all times. Often, a wetland is dry during a period of low river flow or during a low tide. For this reason, only hydric soils and hydrophytic plants should be used as reliable indicators of a wetland.

The broadest classification of wetlands includes two categories: freshwater and saltwater wetlands. Freshwater wetlands occur inland at the edges of rivers, streams, lakes, and other depressions that regularly fill with rainwater. Saltwater wetlands occur along the coast in bays, where salt water and fresh water mix and wave energy is reduced.

Freshwater Wetlands

Of the two categories, freshwater wetlands are by far the most common. Freshwater wetlands are subdivided into two categories: tree-dominated types and grass-dominated types. Tree-dominated freshwater wetlands include areas that are frequently covered with water (such as cypress swamps) and those that are only occasionally covered with water (such as bottomland forests). Grass-dominated types include freshwater marshes, prairie potholes, and bogs.

While freshwater marshes are widespread, prairie potholes and bogs occur regionally in the United States. Prairie potholes are located in the central portion of the United States, while bogs are found in the Northeast and Great Lakes regions.

Wetlands are a special ecotone, or transitional environment, with characteristics of both dry and aquatic habitats and therefore are home to a large variety of plants, animals, birds, and other forms of life. (PhotoDisc)

Saltwater Wetlands

Saltwater wetlands are also subdivided into tree-dominated and grass-dominated types. Tree-dominated types include tropical mangrove swamps. Grass-dominated types can be further subdivided into salt marshes and brackish marshes. Salt marshes occur in bays along the coast where salt water and fresh water mix in almost equal proportions. Brackish marshes occur farther inland than salt marshes do; their mix contains less seawater and more fresh water. Both grass-dominated types are common in bays along the Gulf of Mexico and the East Coast of the United States.

The Biota of Wetlands

The most noticeable feature of all wetlands is the abundance of plant life. A variety of plant species thrive in wetlands, but each occurs only in a particular kind of habitat. Freshwater wetlands that are frequently flooded provide a favorable habitat for water-tolerant trees, such as bald cypress and water tupelo, and water-tolerant herbaceous plants, such as cattail, arrowhead, bulrush, spike rush, water lily, and duckweed. Less frequently flooded freshwater areas support trees such as willow, cottonwood, water oak, water hickory, and red maple. Seawater areas in tropical bays favor the development of mangroves, while temperate bays favor the development of cordgrass.

Wetland plants provide a habitat for a variety of animals. Cypress swamps and cattail marshes support a large assortment of animals, including alligators, ducks, crayfish, turtles, fish, frogs, muskrat, wading birds, and snakes. Likewise, mangrove prop roots provide attachment sites for a variety of invertebrates and shelter for numerous small fish, while upper branches provide roosting and nesting sites for birds. In salt marshes, mussels live among cordgrass roots, while snails, fiddler crabs, oysters, and clapper rails live among plant stalks. When water covers cordgrass at high tide, plant stalks shelter small fish, crabs, and shrimp seeking refuge from large predators.

The Value of Wetlands

The amount of plant material produced in wetlands is higher than that produced in most aquatic and terrestrial habitats. This large amount of plant material supports an abundance of animal life, including commercially important species such as crayfish, ducks, fish, muskrat, shrimp, and crabs.

The biotic value of wetlands is well recognized, but it represents only a part of their total value. Wetlands provide "services" for other areas that often go unrecognized. For example, freshwater wetlands are capable of storing large amounts of water during periods of heavy rainfall. This capability can be important in minimizing the impact of flooding downstream. Saltwater wetlands along coastlines are an effective barrier against storms and hurricanes. These natural barriers hold back the force of winds, waves, and storm surges while protecting inland areas. Wetlands are also capable of increasing water quality through the trapping of sediment, uptake of nutrients, and retention of heavy metals. Sediment trapping occurs when moving water is slowed enough by grass and trees to allow suspended sediment particles to settle. Wetland plants take up nutrients, such as nitrates and phosphates, from agricultural runoff and sewage. For this reason, wetlands are used as a final treatment step for domestic sewage from some small cities. Wetland soils are capable of binding heavy metals, effectively removing these toxic materials from the water.

Wetlands Loss and Preservation
It is estimated that the United States once contained more than 200 million acres of wetlands. Less than half this amount remains today. Once considered wastelands, wetlands were prime targets for "improvement." Extensive areas of freshwater wetlands and prairie potholes have been drained and filled for agricultural development. Saltwater wetlands have been replaced by urban or residential development and covered with dredge spoil. Wetlands loss rates have slowed, but an estimated 300,000 acres continue to be lost each year in the United States. The loss of wetlands habitat threatens the survival of a number of animal species, including the whooping crane, American crocodile, Florida panther, manatee, Houston toad, snail kite, and wood stork.

Since the 1970's the rate of wetlands loss has slowed for several reasons. One is the passage of federal and state laws designed to protect wetlands; another is the efforts of conservation organizations. At the federal level, the single most effective tool for wetlands preservation is Section 404 of the Clean Water Act. Section 404 requires that a permit be issued before the release of dredge or fill material into U.S. waters, including wetlands. At the state level, Section 401 of the Clean Water Act allows states to restrict the release of dredge or fill material into wetlands. Subsequent legislation, notably the North American Wetlands Conservation Act of 1989, worked to conserve wetland habitat. The 1989 act was passed in part to support the North American Waterfowl Management Plan, an international agreement between Canada, Mexico, and the United States to protect wetland/upland habitats on which waterfowl and other migratory birds in North America depend. In December, 2002, President George W. Bush signed the North American Wetlands Conservation Reauthorization Act, intended to "keep our water clean and help provide habitat for hundreds of species of wildlife." Several conservation organizations also support wetlands preservation, including Ducks Unlimited, the National Audubon Society, the National Wildlife Federation, and the Nature Conservancy. These organizations keep the public informed regarding wetlands issues and are active in wetlands acquisition.

Steve K. Alexander, updated by Christina J. Moose

See also: Biomes: determinants; Biomes: types; Chaparral; Deserts; Forests; Grasslands and prairies; Habitats and biomes; Lakes and limnology; Marine biomes; Mediterranean scrub; Mountain ecosystems; Old-growth forests; Rain forests; Rain forests and the atmosphere; Rangeland; Reefs; Savannas and deciduous tropical forests; Taiga; Tundra and high-altitude biomes.

Sources for Further Study

Hey, Donald L., and Nancy S. Philippi. *A Case for Wetland Restoration*. New York: Wiley, 1999.

Littlehales, Bates, and William Niering. *Wetlands of North America*. Charlottesville, Va.: Thomasson-Grant, 1991.

Mitchell, John. "Our Disappearing Wetlands." *National Geographic* 182, no. 4 (1992).

Mitsch, William J., and James G. Gosselink. *Wetlands*. 3d ed. New York: John Wiley, 2000.

Monks, Vicki. "The Beauty of Wetlands." *National Wildlife* 34, no. 4 (1996).

Vileisis, Ann. *Discovering the Unknown Landscape: A History of America's Wetlands*. Washington, D.C.: Island Press, 1999.

Watzin, Mary, and James Gosselink. *The Fragile Fringe: Coastal Wetlands of the Continental Unites States*. Rockville, Md.: National Oceanic and Atmospheric Administration, 1992.

WILDLIFE MANAGEMENT

Type of ecology: Restoration and conservation ecology

Wildlife management strives to allow the use of ecological communities for human benefit while preserving their ecological components unharmed. It also seeks to restore biological communities by managing habitats and controlling the taking of organisms for sport or economic gain.

Wildlife management, also known as game management, is often compared with farming or forestry, because one of its goals is to ensure annual "crops" of wild animals. Conservationist Aldo Leopold, in 1933, defined "game management" as the art of making land produce sustained annual crops of wild game for recreational use. At that time, animals considered to be game included deer and animals such as coyotes that do damage to domestic animals or crops. Now, however, the term "wildlife" has replaced game, and virtually all living organisms, including invertebrates and plants, are included in management considerations.

Approaches to Wildlife Management

The process of wildlife management has moved through a sequence of six approaches: the restriction of harvest (by law); predator control; the establishment of refuges, reserves, and parks; the artificial stocking of native species and introduction of exotic ones; environmental controls, or management of habitat; and education of the general public. All six are used in modern wildlife management programs, but most emphasis is placed on habitat management and control of harvest.

All fifty states of the United States have departments responsible for wildlife conservation. An appointed board of directors or commission oversees the actions of the departments. Groups for wildlife law enforcement, research, management, and information and education make recommendations to the board of directors regarding wildlife management actions. The federal government of the United States also has many agencies that manage wildlife on public lands. The U.S. Fish and Wildlife Service is involved with animals that cross state lines, including migratory birds such as waterfowl, marine mammals, and any plants and animals listed as rare or endangered under the National Environmental Protection Act of 1970. Other agencies, such as the U.S. Forest Service, Bureau of Land Management, Soil Conservation Service, and the U.S. National Park Service, do extensive wildlife work. Many private organizations, such as the National Wildlife Federation, the Audubon Society, and the Sierra Club actively promote wildlife conservation.

Wildlife management decisions involve the entire range of biological, sociological, political, and economic considerations of human society. Today, the wildlife resource in the United States is managed primarily either for consumptive use (such as sport hunting) or for nonconsumptive use (such as bird-watching). Virtually all wildlife management problems are related to the large human population of the earth. Some specific problems are habitat loss (for example, the destruction of tropical rain forests), pollution, diseases introduced by domestic animals into wildlife populations, and the illegal killing of animals for their parts, such as the poaching of elephants for their ivory.

Managing Wildlife Communities

A wildlife manager must first determine the physical and biological conditions of the organism or organisms being managed. Issues include what the best habitat for the animal is and how many animals this habitat can support. The stage of ecological succession determines the presence or absence of particular animals in an area. All animals need food, water, and protection from weather and predators. Special needs, such as a hollow tree in which to raise young, for example, must be fulfilled within the animal's home range. Wildlife managers attempt to remove or provide items that are most limiting to a population of animals. In many respects, solving wildlife management problems is an art; it is similar to medicine in that it often must deal with symptoms (birds dying, for example) and imprecise information.

The stage of ecological succession may be maintained by plowing lands, spraying unwanted plants with a chemical to kill them, or using fire, under controlled conditions, to burn an area to improve the habitat for a certain wildlife group. Refuges and preserves may be set aside to assure that some of the needed habitat is available; nest boxes and water supplies may even be provided.

Periodic surveys of the number of animals in a population provide guidelines for their protection. If animals are more abundant than the lowest carrying capacity, a controlled harvest may be allowed. Sustained annual yield assures that no more than the population surplus is taken. Wildlife laws protect the animals, provide for public safety, often set ethical guides for sporting harvest, and attempt to provide all hunters with an equitable chance of obtaining certain animals (for example, by setting bag limits). If proper wildlife management procedures are followed, no animal need become rare or endangered by sport hunting. Market hunting, the taking of animals for the sale of their products, such as meat or hides, has been stopped in the United States since the 1920's and is also illegal in most other areas of the world. There are almost no societies left that are true subsistence hunters—that is, living exclusively on the materials produced by the wildlife resource.

The Need for Wildlife Management
The proper management of wildlife resources, based on sound ecological principles, is essential to the well-being of humans. All domestic plants and animals came from wild stock, and this genetic reservoir must be maintained. Maintaining the web of life that includes these organisms is necessary for human survival. Wildlife resources are used by at least 60 percent of the citizens of the United States each year, and about 6 percent are sport hunters. Wildlife provides considerable commercial value from products, such as meat; it also offers aesthetic values of immeasurable worth. Seeking and observing wildlife provides needed relief from the everyday tensions of human life. Moreover, by observing wildlife reactions to environmental quality, investigators can monitor the status of the biological system within which humans live. Wildlife populations serve as a crucial index of environmental quality.

Perhaps most important of all, wildlife management helps preserve the biodiversity of communities and ecosystems. Without an effort toward including these ecological considerations along with economic resource concerns, habitat and species loss would quickly ensue, even more rapidly than it now does, in response to urban, suburban, industrial, and agricultural development. Wildlife management is a dynamic activity that, to be effective, must reflect an understanding of and respect for the natural world. It cannot be practiced in a vacuum but must encompass the realm of complex human interactions that often have conflicting goals and values. Aldo Leopold once defined conservation as man living in harmony with the land; successful wildlife management will help assure that this occurs.

David L. Chesemore

See also: Biodiversity; Conservation biology; Deforestation; Endangered animal species; Endangered plant species; Erosion and erosion control; Forest management; Genetic diversity; Grazing and overgrazing; Habitats and biomes; Landscape ecology; Multiple-use approach; Old-growth forests; Reforestation; Restoration ecology; Species loss; Sustainable development; Urban and suburban wildlife; Zoos.

Sources for Further Study

Anderson, S. H. *Managing Our Wildlife Resources.* 3d ed. Upper Saddle River, N.J.: Prentice Hall, 1999.

Bailey, James A. *Principles of Wildlife Management.* New York: John Wiley & Sons, 1984.

Bissonette, John A., ed. *Wildlife and Landscape Ecology: Effects of Pattern and Scale.* New York: Springer, 1997.

Cooperrider, Allen Y., R. J. Boyd, H. R. Stuart, and Shirley L. McCulloch. *Inventory and Monitoring of Wildlife Habitat.* Washington, D.C.: U.S. Government Printing Office, 1986.

Dasmann, R. F. *Wildlife Biology.* New York: John Wiley & Sons, 1981.

Di Silvestro, Roger, ed. *Audubon Wildlife Report, 1986.* New York: National Audubon Society, 1986.

Giles, R. H., Jr. *Wildlife Management.* San Francisco: W. H. Freeman, 1978.

Leopold, Aldo. *Game Management.* New York: Charles Scribner's Sons, 1939. Reprint. Madison: University of Wisconsin Press, 1986.

Matthiessen, Peter. *Wildlife in America.* New York: Viking Press, 1987.

Robinson, W. L., and E. G. Bolen. *Wildlife Ecology and Management.* 4th ed. Upper Saddle River, N.J.: Prentice Hall, 1999.

ZOOS

Type of ecology: Restoration and conservation ecology

Keeping wild animals has evolved, over the past five thousand years, from animal collections maintained by ancient societies to modern zoological gardens and aquariums with significant programs in wildlife appreciation, education, science, and conservation. Originally entertainment venues, zoos have shifted their focus to education and active conservation of endangered and threatened species.

Nearly 600 million people worldwide visit a zoo each year—roughly 10 percent of the global population. Modern zoos, often called wildlife conservation parks or natural wildlife parks, have replaced cages of concrete and steel with simulated natural environments, and new animals are obtained via selective breeding instead of being captured from the wild.

History of Zoos

Early zoos were the sole province of the wealthy; the first recorded zoo in history belonged to a Chinese emperor in 1100 B.C.E. It was not until the nineteenth century that zoos were open to the public. The word "zoo" derives from the phrase "zoological park," and that was what the first zoos were designed to be: afternoon diversions along the same lines as the amusement park or the circus. Exotic beasts from newly charted regions were captured and displayed with little regard for their health or emotional well being. Mortality was high, and display animals were constantly replaced with animals captured from the wild, of which there seemed to be an inexhaustible supply.

The first zoo to use moats to separate animals from visitors was established in Germany by Carl Hagenbeck in 1907. These moats provided visitors with an unobstructed view and, depending on their placement, made it seem as if the animals were free. While the bars were gone, the habitat was still nothing like what the animals were accustomed to in the wild. Those animals that did not spend their days sleeping often displayed near-psychotic behavior patterns, such as pacing, head butting, and even self-mutilation.

A New Role

Two things changed the way zoos functioned during the twentieth century. First, movies and television allowed potential visitors to see the animals in their natural habitats, and suddenly giraffes, lions, and zebras were no longer quite so exotic. Second, wild animals were becoming more scarce, and words such as "conservation" and "endangered" entered the collective vocabulary. Acquiring specimens from the wilderness became more costly, and zoos began to look at internal breeding programs to replenish their stock. However, they found that animals kept in unnatural and in some cases inhumane conditions would not breed.

New zoo enclosures were designed to encourage natural behavior in animals by replicating their natural environment as much as possible while still ensuring the safety of both the animals and the zoo visitors. Animals began receiving healthier diets and, when possible, were allowed to feed in much the same way they would in the wild—by digging, foraging, or grazing. Human contact with orphaned and injured animals was kept to an absolute minimum, and some zoos took the additional step of not naming their animals to discourage anthropomorphism. By 1995, 80 percent of the mammals on display in zoos were born in captivity.

The Lake Superior Zoo in Duluth, Michigan, became home to an Alaskan brown bear named Trouble, who repeatedly broke into the zoo at Anchorage, Alaska. Zoos and wildlife preserves are increasingly the only refuge for endangered and otherwise troubled forms of wildlife whose habitats and lives have been seriously compromised by human activity. (AP/Wide World Photos)

Another trend that began in the late twentieth century was the building of "wild animal parks." The San Diego Zoo, for example, established one of the earliest wild animal parks northeast of San Diego near Escondido, California, which turns the tables by restricting human visitors behind fences and within tram cars as the animals roam freely over large land tracts that approximate their natural habitats.

Mission: Preservation

As concern over endangered animal species, coupled with ecologists' alarm over a decline in biodiversity, spread at the end of the twentieth century, zoos increasingly took on the mission of preservation and breeding. The American Zoo and Aquarium Association (AZA), a nonprofit organization dedicated to the advancement of zoos and aquariums in the areas of conservation, education, science, and recreation, states its members' mission as "work[ing] cooperatively to save and protect the wonders of the living natural world." The AZA has accredited more than two hundred zoos worldwide and sponsors a variety of research, education, and preservation programs. One of these, the Species Survival Plan (SSP) program, was established in 1981 to "help ensure the survival of selected wildlife species into the future and to provide a link between zoo and aquarium animals and the conservation of their wild counterparts" through breeding and, where possible, reintroduction into the wild.

Zoo managers continue to struggle to balance science, conservation biology, scarce resource allocation, and ethics. Among the choices that must be made are whether predators should be offered the chance to exercise natural hunting behaviors by being offered live prey, or whether zoos should maintain potentially deadly animals that are necessary for breeding programs but are dangerous and difficult to control, such as macaques, many of which harbor the deadly hepatitis B virus, or adult male elephants. Another dilemma is the question of what should become of "surplus" animals that are inbred, unable to reproduce, or are otherwise genetically inferior.

Municipal bureaucracies can also hamper zoo conservation efforts. Zoo managers must often combat local governments and public opinion when dealing with unpopular issues, such as surplus animals and resource allocation. In addition, budget cuts have forced zoo managers to turn to the private sector for financial assistance. Fund-raising activities range from the traditional "adopt an animal" programs to the extraordinary commercial venture of selling "exotic compost."

There are those who question whether zoos should exist at all—

whether it is cruel or unusual to take animals from their natural habitat and place them on display. The People's Republic of China, for example, has "rented" its zoo's giant pandas when the pandas might be better served by remaining in the wild or in a captive-breeding program that would allow them to replenish their numbers. Critics claim that the money devoted to zoos and captive-breeding programs would be better spent on preserving the animals' natural habitats. To combat these types of criticism, some zoos began to change their focus from "collecting" wildlife to "protecting" wildlife, also known as field conservation. In these new exhibits, zoo visitors view exhibits linked with protection and conservation programs in natural habitats, allowing visitors to connect what they are seeing in captivity to what's worth saving in the wilderness. Some zoos have taken the additional step of "adopting" wildlife refuges.

Despite the criticism, the fact remains that there are many animal species that could not survive without the existence of zoos and captive-breeding programs. Ironically, where historical zoos replenished their stock from the wilderness, some zoos are now replenishing the wilderness with captive-bred animals.

P. S. Ramsey, updated by Christina J. Moose

See also: Biodiversity; Conservation biology; Deforestation; Endangered animal species; Genetic diversity; Restoration ecology; Species loss; Sustainable development; Urban and suburban wildlife; Wildlife management.

Sources for Further Study

Bell, Catharine, et al., eds. *Encyclopedia of the World's Zoos.* Chicago: Fitzroy Dearborn, 2001.

Croke, Vicki. *The Modern Ark: The Story of Zoos, Past, Present, and Future.* New York: Charles Scribner's Sons, 1997.

Hoage, R. J., and William A. Deiss, eds. *New Worlds, New Animals: From Menagerie to Zoological Park in the Nineteenth Century.* Baltimore: The Johns Hopkins University Press, 1996.

Kisling, Vernon N., Jr., ed. *Zoo and Aquarium History: Ancient Animal Collections to Zoological Gardens.* Boca Raton, Fla.: CRC Press, 2001.

Koebner, Linda. *Zoo Book: The Evolution of Wildlife Conservation Centers.* Preface by William Conway. New York: T. Doherty, 1994.

Norton, Bryan G., Michael Hutchins, Elizabeth F. Stevens, and Terry L. Maple, eds. *Ethics on the Ark: Zoos, Animal Welfare, and Wildlife Conservation.* Washington, D.C.: Smithsonian Institution Press, 1995.

Tudge, Colin. *Last Animals at the Zoo: How Mass Extinction Can Be Stopped.* Washington, D.C.: Island Press, 1992.

Wemmer, Christen M., ed. *The Ark Evolving: Zoos and Aquariums in Transition.* Front Royal, Va.: Smithsonian Institution Conservation and Research Center, 1995.

GLOSSARY

abiotic: Not living; used to refer to the nonliving elements of an ecosystem or biome, consisting of climate, the minerals in soil, rocks, water, oxygen, carbon dioxide, and other physical components.

abundance: The density and prevalence of organisms living in a particular population, community, or ecosystem.

abyssal marine zone: The dark marine zone that extends from the ocean floor to where the continental slope begins.

acclimatization: A process by which animals habituate their physiological responses to the conditions of a particular environment.

acid deposition: The process whereby acidic (very low pH, usually under 5.6) gases, particles, and precipitation (rain, fog, dew, snow, or sleet) fall on the surface of the earth.

active or wide foraging: Moving across a relatively large area in search of prey.

adaptation: In evolutionary biology, a heritable structure, physiological process, or behavioral pattern that gives an organism a better chance of surviving and reproducing; in physiology, the change in the response of a sense organ following continuous application of a constant stimulus.

adaptive radiation: The relatively rapid evolution of several new species from a common ancestor following invasion of a new geographic region or ecological niche, or exploitation of a new ecological opportunity.

aerobic: Characterizing any biological process that occurs in the presence of free oxygen.

aestivation: *See* estivation.

aggregation: A group of organisms that live in closer proximity than they would were they to be randomly or evenly distributed. Aggregations of the same species are known as populations.

aggression: A physical act or threat of action by one individual that reduces the freedom or genetic fitness of another.

agricultural ecology: Also called agroecology, the study of agricultural ecosystems, their components (such as crop species), functions, interactions, and impact on natural ecosystems and abiotic factors such as atmospheric and water systems—often with an emphasis on the development of sustainable systems.

agricultural revolution: The transition by humans from hunting and gathering all their food to domesticating plants for food.

alleles: Different forms of a particular gene, located at the same position on a chromosome. Most genes have at least two naturally occurring alleles. These alleles may be the same or may be different.

allelochemical: A general term for a chemical used as a messenger between members of different species; allomones and kairomones are allelochemicals, but hormones and pheromones are not.

allelopathy: The phenomenon in which an organism produces and releases chemicals that are toxic to, or inhibit the growth of, another organism, although sometimes with beneficial effects to the other organism.

allopatric speciation: The evolution of two new species as a result of the separation of two groups of the same species from each other.

allopolyploidy: Formation of a new species resulting from the mating of two different species, resulting in a sterile or reproductively isolated hybrid.

alpine tundra: The area of Europe, Asia, and North America north of the boreal coniferous forest zone, where the soils remain frozen most of the year, constituting about 3 percent of the earth's surface.

altruism: A behavior that increases the fitness of the recipient individual while decreasing the fitness of the performing individual.

amensalism: An interspecies relationship in which one species is harmed while the other remains uneffected.

anaerobic: Lacking, or living in the absence of, oxygen.

antagonism: Any type of interactive, interdependent relationship between two or more organisms that is destructive to one of the participants.

antithesis, principle of: The observation that signals communicating opposite meaning tend to be expressed using displays having opposite characteristics.

aposematic coloration: Bright warning coloration that toxic species use to advertise their distastefulness to would-be predators.

aquatic ecology: The study of the ecology of freshwater systems (rivers, lakes), estuaries, and marine environments (both coastal and open ocean), including the physical, chemical, and biological processes associated with them.

Arctic tundra: Treeless biome of very cold climates near to and north of the Arctic Circle, in which the predominant plants are low-growing, perennial woody plants and grasses. Lichens and mosses may also be common.

area-sensitive species: Species that require large blocks of natural area for activities such as reproduction, finding food, and raising of young, are called area-sensitive species. This required area may include portions of open habitat, edge, and interior habitat. Area-sensitive species are especially sensitive to any reduction in area caused by habitat fragmentation.

artificial selection: Choices made by plant breeders to produce varieties of plants that have some desirable quality, such as improved yield, greater height, or an unusual flower color.

asexual reproduction: Reproduction of cells or organisms without the transfer or reassortment of genetic material, resulting in offspring that are genetically identical to the parent.

autoecology: *See* physiological ecology.

autopolyploidy: Formation of a new species by the doubling of chromosomes of a single existing species. Many related species of plants within a genus have been found to result from repeated occurrences of autopolyploidy.

autotrophs: Organisms that have the ability to make their own food from inorganic substances. *See also* photosynthesis; primary producers.

back-cross: A cross involving an offspring individual crossed with one of its parents.

banding: Technique for studying the movement, survival, and behavior of birds by means of identification tags.

Batesian mimicry: An evolutionary trend in which an edible species mimics the form of a distasteful species to avoid predation.

behavioral ecology: The systematic study of the strategies animals use to overcome environmental problems and the adaptive value of those strategies.

benthos: The area of the ocean floor; organisms associated with the sea bottom.

bioaccumulation: *See* biomagnification.

bioclimatic zone: A zone of transition between differing yet adjacent ecological systems.

biodegradable: Capable of being broken down, or degraded, into simpler substances by natural decomposers.

biodiversity: This term represents an amalgamation of biology and diversity. Ecologists typically recognize three types of biodiversity—species biodiversity, genetic biodiversity, and habitat biodiversity. Habitat biodiversity refers to the variety of habitats in a given landscape, genetic biodiversity refers to the number of alleles in a species genome or gene pool in a given area. Species diversity refers to both the variety of species and equal numbers of individuals among each species.

biogenetic law: Ernst Haeckel's term for his generalization that the ontogeny of an organism recapitulates the adult stages of its ancestors (recapitulation).

biogeochemical cycles: Movement of elements or water through both living and nonliving parts of an ecosystem. Carbon as carbon dioxide is made into carbohydrate during photosynthesis and released through decay to the nonliving atmosphere, from which it can later be reused in photosynthesis.

biogeography: The science that seeks to understand spatial patterns of biodiversity. *See also* island biogeography.

bioluminescence: Production of visible light by living organisms.

biomagnification: Also called bioaccumulation, the increasing accumulation of a toxic substance in progressively higher feeding levels

biomass: The weight of organic matter in an environment or ecosystem, often expressed in terms of grams per square meter per year.

biomes: The primary, large-scale ecosystems of the world, largely identified with geographical regions typified by climate and weather and classified in the Köppen system on the basis of precipitation, temperature, climate, soil types, flora, fauna, and location. Major biomes include Tropical (rain forests, savannahs, tropical deciduous forests, and tropic scrub), Mid-Latitude to Equatorial (temperate, hot, and cold grasslands, including steppe and chapparal; hot and cold deserts), Continental (woodlands, deciduous temperate forests, mediterranean woodland and shrub), Moderate Continental (moderate grasslands, deciduous forests, taiga and boreal forests), and Polar (ice caps, tundra).

biopesticides: Biological agents, such as viruses, bacteria, fungi, mites, and other organisms used to control insect and weed pests in an environmentally and ecologically friendly manner.

biosphere: Specifically, the 20-kilometer-thick zone extending from the floor of the oceans to the top of mountains, within which all life on earth exists. Generally, the sum of all the occupiable habitats for life on earth.

biota: All living things in a particular area, including microbes, fungi, algae, plant life, and animals.

biotechnology: Combination of techniques whereby humans are able to alter permanently the genetic makeup of organisms. Includes the industrial application of these techniques.

biotic: Living. Refers to the living components of an ecosystem or biome, consisting of all organisms.

boreal forest: Located in two broad belts of vegetation that stretch from east to west in the Northern Hemisphere, the primarily coniferous forests that dominate this biome, which is also known as taiga.

bottleneck effect: In evolution, the reduction in size of a population causing a major loss of genetic variation. If the population size later expands, the new larger population will be genetically uniform and may lack the ability to survive in a changing climate.

brood: All the immature insects within an insect colony, including eggs, larvae, and, in the *Hymenoptera*, the pupal stage; also, to cover young with the wings.

brood parasite: *See* nest parasite.

browser: An animal that feeds on leaves and twigs from trees.

budding: A form of asexual reproduction that begins as an outpocketing of the parental body, resulting in either separation from or continued connection with the parent, forming a colony.

C_3 plants: Plants whose system of photosynthesis produces a three-carbon compound as the first identified compound after the uptake of carbon dioxide during the light-independent reactions. *See also* CAM plants.

C_4 plants: Plants whose system of photosynthesis produces a four-carbon compound as the first identified compound after the uptake of carbon dioxide during the light-independent reactions. C_4 photosynthesis is distinguished from CAM photosynthesis because C_4 occurs during the day and CAM occurs during the night. C_4 plants are especially adapted to hot, dry climates. Corn is an example. *See also* CAM plants.

calorie: The traditional unit of heat; one calorie is the amount of heat required to raise the temperature of one gram of water 1 degree Celsius.

CAM plants: Plants in desert biomes that use a crassulacean acid metabolism to take in carbon dioxide during the night and store it as an acid, and then use the carbon dioxide in the light-independent reactions during the day, when sunlight is available. Cacti are CAM plants. *See also* C_3 plants; C_4 plants.

Cambrian explosion: The main period of evolutionary expansion in the Phanerozoic era at the base of the Cambrian period, 544 million years ago, which marks the development of all the modern phyla of organisms.

camouflage: Patterns, colors, and/or shapes that make it difficult to differentiate an organism from its surroundings.

canopy: The uppermost portion of a rain forest, which shades the understory and the forest floor.

capture-recapture: *See* mark-capture-release method.

carbohydrates: Large class of organic molecules containing starch, carbon, hydrogen, and oxygen and in which the ratio of hydrogen to oxygen is two to one, the same as in a molecule of water. Sugars and cellulose are examples.

carbon cycle: Biogeochemical cycle of the element carbon.

carbon fixation (CO_2): Process by which carbon dioxide is made into glucose during photosynthesis. This occurs during the part of photosynthesis called the Calvin cycle.

carcinogen: Any physical or chemical cancer-causing agent.

carnivore: A member of the meat-eating order Carnivora, which includes dogs, cats, weasels, bears, and their relatives.

carnivorous plant: Plant that traps insects and digests them. These plants usually live in nitrogen-poor habitats and use the insect proteins to supplement their nitrogen intake.

carnivory: Subsisting or feeding on meat or flesh.

carrion: Dead animal flesh.

carrying capacity: The maximum number of animals that a given area can support indefinitely.

caste: One of the recognizable types of individuals within an insect colony, such as queens, workers, soldiers, and males or drones; usually these individuals are physically and behaviorally adapted to perform specific tasks.

catastrophism: A scientific theory which postulates that the geological features of the earth and life thereon have been drastically affected by natural disasters of huge proportions in past ages.

cellulose: A fibrous polysaccharide that chiefly constitutes the cell walls of plants and is not easily dissolved in water. Primary consumers, including ruminants, require specialized digestive systems to break down cellulose.

census: The counting of populations of naturally occurring organisms to understand their ecology more fully.

CFCs: *See* Chlorofluorocarbons (CFCs).

chaparral: Biome found along the coast of Southern California, characterized by short trees with leathery leaves, shrubs, and open grassy areas.

character displacement: A change in the morphological, behavioral, or physiological state of a species without geographical isolation, as a result of natural selection arising from competition between one or more ecologically similar species.

chemical ecology: Ecology that concerns the biochemicals (called semiochemicals) produced and released by organisms that have physiological and behavioral effects on other organisms. Studies in chemical ecology are often interdisciplinary (integrating several fields of science such as chemistry and ecology) but also more specific areas in biochemistry (such as biosynthesis of compounds), molecular biology, or physiology (reception of the compounds and transmission of nerve impulses), as well as in behavioral ecology (orientation movements of an organism) and population ecology (aggregation and competition of organisms) and even the interactions among trophic levels (such as predator-prey interactions). Evolutionary studies at all these levels are of interest to understand how stable the semiochemical systems are and whether adaptations to new systems are constrained.

chemical pollutants: Harmful chemicals manufactured and released to the environment.

chemosynthetic autotrophs: Organisms (usually bacteria) that make complex food molecules from simpler molecules using energy of chemical reactions rather than light energy used by photosynthetic autotrophs.

chlorofluorocarbons (CFCs): A group of very stable compounds used widely since their development in 1928 for refrigeration, coolants, aerosol spray propellants, and other uses; once risen in stratosphere, they cause ozone depletion.

circadian rhythm: A physiological or behavioral cycle that occurs roughly in a twenty-four-hour pattern.

cladistics: System of describing evolutionary relationships in which only two groups, or clades, branch from each ancestral group. The more recently two clades diverged, the more characteristics they have in common and the closer they will appear in a cladistic diagram.

class: The taxonomic category composed of related genera; closely related classes form a phylum or division.

classification: The arranging of organisms into related groups based on specific relationships. *See also* systematics; taxonomy.

clear-cutting: The removal of all trees from an area.

climax community: Group of plants that appear late in succession and are not replaced by plants of different species unless the climate changes or the area is disturbed, as by fire or cultivation.

cline: A graduated series of populations of the same species. Each population has a slightly different physiology from the ones on each side. Clines typically develop where environmental factors change in a gradual way, such as from the bottom of a mountain to the top.

clone: An organism that is genetically identical to the original organism from which it was derived.

cloning: The technique of making a perfect genetic copy of a DNA molecule, a cell, or an entire organism.

clutch: A group of eggs laid in a single reproductive effort.

coevolution: Simultaneous evolutionary change through time of two species, such as a flower and its pollinating insect, each influenced by the changes the other species is undergoing. Over a period of time, the two species often become dependent on each other, so that one would not survive the disappearance of the other.

cognitive ethology: The study of animal intelligence.

cohort: A group of organisms of the same species, and usually of the same population, that are born at about the same time.

colony: A cluster of genetically identical individuals formed asexually from a single individual.

coloration: *See* aposematic coloration; cryptic coloration.

commensalism: A type of coevolved symbiotic relationship between different species that live intimately with one another without injury to any participant.

communication: The exchange of information between members of a species by means of chemical signals (pheromones), displays, calls, and other means.

community: A population of plants and animals that live together and interact with one another through the processes of competition, predation, parasitism, and mutualism, making up the biotic part of an ecosystem.

community ecology: The study of the impacts that populations of different species have on populations of other species with which they interact, be those interactions between plants and other plants, animals and other animals, or plants and animals. The emphasis is on how these populations of different species change, enhance, or delimit one another. Population ecology is related but is focused on the growth and change in populations of discrete species.

comparative physiology: *See* physiological ecology.

compartmentalization: A characteristic of most communities, in which a given set of producers tends to be consumed by a limited number of consumers, which in turn are preyed upon by a smaller number of predators, and so on.

competition: The interactions among individuals that attempt to utilize the same limited resource.

competitive exclusion, principle of (Gause's principle): If two or more species compete for the same niche, one of them will be successful, and the other will be eliminated over time.

coniferous forest: Large group of trees that are predominantly conifers, such as a pine forest or a spruce-fir forest.

conjugation: Type of sexual reproduction that occurs in green algae such as *Spirogyra* and in certain kinds of fungi. Also, the transfer of genetic material from one bacterium to another through a cytoplasmic bridge.

connectivity: The ability of organisms to use corridors of habitat to disperse from one habitat patch to another.

conservation biology: The use of biological science to design and implement methods to ensure the survival of species, ecosystems, and ecological processes. Conservation biologists are concerned with the process of speciation, the measurement of biodiversity, and factors involved in the extinction process. However, the primary thrust of their efforts is the development of strategies to preserve biodiversity; hence, conservation biology is a value-laden science.

conservation easement: An arrangement in which a national government, private organization, or consortium of countries compensates a tropical country for protecting a specific habitat. *See also* debt-for-nature swap.

consort pair: A temporarily bonded pair within a polygamous group; also called consortship.

conspecific: A member of the same species.

constriction: A method of killing prey using increasingly tight coils around the body to trigger stress-induced cardiac arrest.

consumer: An organism other than a primary producer—that is, one that eats other organisms, including primary consumers (herbivores), secondary consumers (omnivores and carnivores), scavengers, and decomposers.

continuous growth: Growth in a population in which reproduction takes place at any time during the year rather than during specific time intervals.

convergent evolution: The process by which evolutionarily unrelated animals tend to resemble one another as a result of adaptations to similar environments.

cooperation: A social behavior in which members of a group act for the good of all.

coral reef: A reef built primarily by coral species.

core species: A species that utilizes the interior area or core of a habitat. Core species such as many neotropical migrants typically require large blocks of contiguous natural habitat for reproduction and are typically very sensitive to the ratio of core-to-edge habitat in a given patch or parcel of landscape.

corridor: Narrow link of habitat that connects two or more patches in a landscape. Natural corridors serve as wildlife dispersal conduits, permitting movement between natural habitats with minimal risk of exposure.

countershading: A form of crypsis involving dark coloration on top and light coloration on the underside.

coupled oscillations: In predator-prey relationships, the waxing and waning of population sizes of two species in a community, based on cycles of predation.

crassulacean acid metabolism (CAM): *See* CAM plants.

crepuscular: Active after sunset and in early morning.

Cretaceous-Tertiary (KT) event: An event that occurred about 66.4 million years ago, sometimes hypothesized to be a meteoritic impact, that marks the boundary between the Cretaceous and Tertiary periods and that initiated the mass extinctions of many species, notably the dinosaurs.

critical period: A very brief period of time in the development of an animal during which certain experiences must be undergone; the effects of such experiences are permanent.

critical photoperiod: Specific day length necessary to produce flowers in long-day and short-day plants.

cross-pollination: The transfer of pollen grains and their enclosed sperm cells from the male portion of a flower to a female portion of another flower within the same species.

crown: The branched, leafy part of a tree.

crypsis: The phenomenon of hiding or remaining hidden.

cryptic coloration: Any color pattern that blends into the background.

cuckold: A partnered male who is helping his mate to raise offspring which are not genetically his own.

cud: Food regurgitated and chewed a second time after its initial ingestion.

cultural diversity: Variety of learned behaviors among individuals of a species.

cyanobacteria: Photosynthetic prokaryotes that were once called blue-green algae.

debt-for-nature swap: An arrangement in which tropical countries act as custodians of the tropical forest in exchange for foreign aid or relief from debt. *See also* conservation easement.

deciduous tropical forest: Forests of tropical regions that shed their leaves during annual dry periods.

decomposers: Bacteria and fungi that break down dead organic matter. In the process, they obtain energy and facilitate the recycling of elements.

deep ecology: A philosophy, introduced by the Norwegian philosopher Arne Naess, that values the natural world and biodiversity in and of themselves and advocates seeing nature as more than a resource for human use.

defense mechanism: Any of a variety of chemical, behavioral, anatomical, or physiological means by which an organism prevents or discourages predation.

definitive host: The host in which a symbiont (the organism living within the host) matures and reproduces.

defoliant: A chemical that kills the leaves of trees.

deforestation: The removal of trees from forests to an extent that degrades the forest biome without human intervention. *See also* reforestation.

deme: A local population of closely related living organisms.

demography: The study of the numbers of organisms born in a population within a certain time period, the rate at which they survive to various ages, and the number of offspring that they produce. Also referred to as demographics.

dendrochronology: The examination and comparison of growth rings in both living and aged woods to draw inferences about past ecosystems and environmental conditions.

dendroclimatology: The study of tree-ring growth as an indicator of past climates.

denitrification: Process in which bacteria convert nitrogenous compounds in the soil to nitrogen gas.

denning: The period of winter sleep during which a bear does not eat, drink, urinate, or defecate.

density: The number of animals present per unit of area being sampled; for example, ten mice per hectare or five moose per square kilometer.

density-dependent growth: Growth in a population in which the per capita rates of birth and death are scaled by the total number of individuals in the population.

density-dependent population regulation: The regulation of population size by factors or interactions intrinsic to the population; the strength of regulation increases as population size increases.

deoxyribonucleic acid (DNA): The genetic material of cells, having the molecular form of a twisted double helix that is linked by purine and pyrimidine base pairs; carries the inherited traits and controls for cell activities.

desert: Biome that receives less than 10 inches of precipitation per year.

desertification: The degradation of arid, semiarid, and dry, subhumid lands as a result of human activities or climatic variations, such as a prolonged drought.

despotism: A type of hierarchy in which one individual rules over all other members of the group and no rank distinctions are made among the subordinates.

detritus feeders (detritivores): An array of small and often unnoticed animals and protists that live off the refuse of other living beings, such as molted shells and skeletons, fallen leaves, wastes, and dead bodies.

diapause: A resting phase in which metabolic activity is low and adverse conditions can be tolerated; also, an interruption in embryonic development.

diatom: Microscopic algae that produce a frustule (a kind of shell) made of silica glass that is highly resistant to weathering. A major type of phytoplankton.

differentiation: The process during development by which cells obtain their unique structure and function.

digestion: The process by which larger organic nutrients are broken down to smaller molecules in the lumen of the gut.

dilution effects: The reduction in per capita probability of death from a predator due to the presence of other group members.

dimorphism: Existence of two distinct forms within a species.

dinoflagellates: A unicellular, mobile type of phytoplankton responsible for deadly algal blooms called red tides.

dioecious: Having two separate sexes, namely male and female.

diploid: Having two sets of chromosomes, usually one derived from the father and one derived from the mother; the normal condition of all cells except reproductive ones. *See also* haploid.

discrete growth: Growth in a population that undergoes reproduction at specific time intervals.

discrete signals: Signals that are always given in the same way and indicate only the presence or absence of a particular condition or state.

disease ecology: Ecological factors influencing, and influenced by, the emergence and spread of infectious diseases, including both plant and animal populations. Considers such questions as how biodiversity affects the spread of disease, patterns of disease spread through populations, and the roles of evolution and genetics.

disharmonic: Ecologically unbalanced.

disjunct: Pertaining to the geographic distribution pattern in which two closely related groups are widely separated by areas that are devoid of either group.

dispersal: The movement of organisms from one geographic area to another; movements may be the result of an animal's own efforts (active dispersal) or the consequence of being transported by natural or human-mediated means (passive dispersal) and can be limited by physical barriers.

dispersion: The pattern or arrangement of members of a population in a habitat; also, the transport or movement of seeds across a substrate such as soil prior to germination.

disphotic marine zone: A transitional marine region between the photic and aphotic zones that may extend to depths of one thousand meters; also called the mesopelagic zone.

display: A social signal, particularly a visual signal, exchanged between animals.

disruptive coloration: Use of stripes, spots, or blotches to break up the body outline and blend into a complex background.

diurnal: Awake and functional during the daylight hours.

divergence: The evolution of increasing morphological differences between an ancestral species and offshoot species caused by differing adaptive pressures.

diversity: The number of taxa (classification groups) associated with a particular place and time. *See also* biodiversity.

DNA: *See* deoxyribonucleic acid.

domestication: A process by which animals are adapted biologically and behaviorally to a domestic (human) environment in order to tame and manipulate them for the benefit of humans.

dominance: The physical control of some members of a group by other members, initiated and sustained by hostile behavior of a direct, subtle, or indirect nature.

dominance hierarchy: A social system, usually determined by aggressive interactions, in which individuals can be ranked in terms of their access to resources or mates.

dominant species: A species in a community that acts to control the abundance of its competitors because of its large size, extended life span, or ability to acquire and hold resources.

dormancy: A period of inactivity that allows a plant or animal to survive unfavorable cold or dryness (hypernation, estivation).

drone: A fertile male social insect.

drumming: Type of nonvocal communication that a bird produces by banging its bill on a hollow tree trunk or other noise-producing object.

dry tropical biomes: The savanna and deciduous tropic forest biomes, which often occur where the annual rainfall is less than that of savannas, most often found in Africa, South America, and Australia.

dynamic equilibrium: Characterizing a community in equilibrium that is always responding to the last disturbance.

dynamically fragile: Characterizing a community that exhibits resilience only within a narrow range of conditions.

dynamically robust: Characterizing a communitiy that exhibits resilience over a wide range of conditions and scales of disturbance.

eclipse plumage: The drab plumage of male birds following the post-breeding molt, in which their bright courtship feathers are replaced by dull earthy feathers that provide inconspicuous coloring.

ecocentrism: A philosophy that emphasizes the value of nature as a whole and an identification of the self with the natural world.

ecoenergetics: The flow of energy through ecological systems at all levels, from individual organisms, populations, and communities to ecosystems and the global environment. Includes abiotic factors (such as geochemical cycles) as well as biotic factors.

ecofeminism: A consideration of gender differences in the experience of the self and nature, which includes an analysis of the tie between the oppression of women and nature.

ecological niche: *See* niche.

ecology: The study of the interactions between organisms and the living (biotic) and nonliving (abiotic) components of their environment, including the distribution and abundance of organisms.

ecomorph: Species of different phyletic origins (at most distantly related) with similar structural and behavioral adaptations to similar niches.

ecophysiology: *See* physiological ecology.

ecosystem: A biological community and the physical environment contained in it.

ecosystem diversity: Variety of biomes and habitats occuring in the biosphere.

ecosystem ecology: The study of the flow of energy into, through, and out of large-scale systems, and how that flow influences all abiotic factors and living organisms in the ecosystem.

ecotone: The meeting place between the two biomes, a transition zone.

ecotourism: Tourism based on the promotion of ecologically important sites, such as national parks, wildlife refuges, endangered ecosystems, and the organisms for which they form the habitat.

ecotoxicology: The study of natural and man-made pollutants and their toxic effects on organisms, populations, communities, and ecosystems, as well as the ways these pollutants impact ecological processes to change ecosystems and their components.

ectoparasite: A parasitic organism that attaches to the host on the exterior of the body.

edge species: Certain species of wildlife that exploit edge habitat, which is the transition habitat zone between two distinct habitats.

elfin forest: A stunted forest growing at high elevations in warm, moist climates.

emergent vegetation: Vegetation that grows tall enough to be visible at the highest level of a rain forest or, in aquatic systems, above the water line.

emigration: The movement of animals out of an area; one-way movement from a habitat type.

encounter effects: The reduction in the probability of death from a predator due to a single group of N members being more difficult to locate than an equal number of solitary individuals.

endangered species: A species of animal or plant, as designated by bodies such as the U.S. Fish and Wildlife Service, that is threatened with extinction.

endemic: Belonging to or native to a particular place; often referring to species that have evolved in a given area and are found nowhere else in the world.

endosymbiosis: A symbiotic relationship in which one member lives inside the other's body.

endocannibalism: A form of human cannibalism in which members of a related group eat their own dead.

endoparasite: A parasitic organism that attaches to an interior portion of the host's body.

endophyte: An organism living within a plant, such as a fungus living in a root.

endosymbiotic theory: Theory that chloroplasts and mitochondria developed from bacteria that moved into and became essential to the survival of early eukaryotic cells.

endotherm: An animal that, by its own metabolism, maintains a constant body temperature (warm-blooded); birds and mammals are endotherms.

energy: The ability to do work. Energy takes several forms, such as chemical energy in the bonds of a compound, light energy, and kinetic energy, the energy of movement.

energy budget: The amount of resources available to an organism, which is accepted to be limited or finite. Energy acquired from food (animals) or sunlight (plants) must be partitioned among growth, maintenance, and reproduction. The greater the energy allocated to the care of offspring, for example, the fewer the offspring that can be produced.

energy flow: The capture of radiant energy, its transformation into chemical energy by producers via photosynthesis, and its translocation through all biological systems via consumers and decomposers. All organisms are considered as potential sources of energy. *See also* food chain.

energy pyramid: A graphical representation of the energy contained in succeeding trophic levels, with maximum energy at the base (producers) and steadily diminishing amounts at higher levels.

entrainment: The synchronization of one biological rhythm to another rhythm, such as the twenty-four-hour rhythm of a light-dark cycle.

entropy: The tendency of complex molecules and other structures to lose energy and become degraded into simpler forms.

environment: All the external conditions that affect an organism or other specified system during its lifetime.

environmental constraints: The physical demands placed upon any species by its surroundings that ultimately determine the success or failure of its adaptations and consequently its success as a species; also called pressures.

epifauna: Animals that live on the sea floor.

epiphyte: A plant that lives nonparasitically on another plant. Usually, epiphytes are not rooted in the ground and are typically found in habitats with high humidity. Spanish moss is an epiphyte.

epizoites: Commensals that live on the skin, fur, scales, or feathers of their hosts.

estivation: Similar to hibernation, a period of reduced activity or dormancy triggered by dry or hot environmental conditions. *See also* hibernation.

estuarine ecosystem: An aquatic ecosystem that occurs where a freshwater river meets a saltwater or ocean environment.

ethology: The study of an animal's behavior in its natural habitat.

euphotic zone: Also called the photic zone, the region of a body of water that is penetrated by sunlight.

eusocial: Characterizing a social system with a single breeding female; other members of the colony are organized into specialized classes (exemplified by bees, ants, and termites).

eutrophication: The overenrichment of water by nutrients, causing excessive plant growth and stagnation, which in turn leads to the death of fish and other aquatic life.

evapotranspiration: The loss of water by means of both evaporation from soil and transpiration from plants.

evergreen: Plant that keeps its leaves during adverse climatic conditions, such as cold temperatures. Examples are pine trees and rhododendrons.

evolution: A process, guided by natural selection, that changes a population's genetic composition and results in adaptations.

evolutionarily stable strategy: A behavioral strategy that will persist in a population because alternative strategies, in the context of that population, will be less successful.

evolutionary ecology: The study of how evolutionary processes such as selection and adaptation influence the interactions of organisms with their environments and shape species and ecosystems.

exogenous: Originating outside an ecosystem, community, population or organism.

exotic species: Organisms that are not naturally found in a place but have been artificially introduced, whether by accident or intentionally. *See also* invasive species.

exponential growth: A pattern of population growth in which the rate of increase becomes progressively larger over time.

extant: Alive and reproducing, as opposed to extinct species.

external fertilization: The union of eggs and sperm in the environment, rather than in the female's body.

extinct: No longer living on earth.

F1 generation: First filial generation; offspring produced from a mating of P generation individuals.

F2 generation: Second filial generation; offspring produced from a mating of F1 generation individuals.

facultative anaerobe: Organism that can survive in an oxygen-poor environment when necessary, using the small amount of energy available from fermentation.

fecundity: The number of offspring produced by an individual.

female: An organism that produces eggs, the larger of two different types of gametes.

fermentation: Set of reactions that change glucose into alcohol and carbon dioxide. A small amount of energy is produced, some of which is captured as ATP.

fertilization (agriculture): Addition of minerals or decayed organic matter to soil that has been depleted of nutrients by farming or other means.

fertilization (sexual reproduction): The fusion of sperm and egg to form a fertilized egg (zygote).

field capacity: Water remaining in soil following rain or irrigation after excess water has been drained off by gravity.

fire climax ecosystem: An ecosystem that depends on periodic fires to clear underbrush; the seeds of many plants in such an ecosystem require fire in order to germinate.

fire ecology: The study of how both natural and planned fires impact ecosystems such as forests.

fitness: The ability of an organism to produce offspring that, in turn, can reproduce successfully; the fitness of organisms increases as a result of natural selection.

fixation: Process by which a gas has been converted to another type of molecule, usually organic.

fledgling period: Period after hatching, during which a nestling grows flight feathers and learns to fly.

food chain: A paradigm representing the links between organisms, each of which eats and is eaten by another.

food pyramid: A diagram representing organisms of a particular type that can be supported at each trophic level from a given input of solar energy in food chains and food webs.

food web: A network of interconnecting food chains representing the food relationships in a community.

foraging: *See* active or wide foraging; sit-and-wait foraging.

fossil: A remnant, impression, or trace of an animal or plant of a past geological age that has been preserved in the earth's crust.

fossil fuel: A fuel product originating from the partial or complete decomposition of carbon-based life-forms (plant and animal remains and fossils) exposed to heat and pressure in the earth's crust over thousands and millions of years. Fossil fuels include crude oil, coal, natural gas, and gasoline.

fossil record: The evolutionary information contained in the fossils found in the earth's crust when compared with the geologic record.

frequency-dependent predation: Predation on whichever species is most common in a community; a frequency-dependent predator will switch prey if necessary.

frozen zoo: A frozen tissue bank, maintained at many zoos, that contains wild animal tissue and reproductive samples for use in future breeding programs.

functional response: The rate at which an individual predator consumes prey, dependent upon the abundance of that prey in a habitat.

fusion-fission community: A society whose members are of both sexes and all ages, which can form and dissolve subgroupings.

gamete: A haploid reproductive cell, usually a sperm or egg.

Gause's principle: *See* competitive exclusion, principle of.

gene: A portion of a DNA molecule containing the genetic information necessary to produce a molecule of messenger RNA (via the process of transcription) that can then be used to produce a protein (via the process of translation).

gene flow: The movement of genes from one population to another through migration and hybridization between individuals belonging to adjacent populations.

gene frequency: The occurrence of a particular allele present in a population, expressed as a percentage of the total number of alleles present.

gene (point) mutation: A change within the hereditary material of a single gene. These tiny mutations cannot be observed by inspecting pictures of chromosomes.

gene pool: All the alleles of all the genes present in a population. There is no limit to the number of alleles in a gene pool; however, an individual may not possess more than two different alleles.

generalized: Not specifically adapted to any given environment.

genetic diversity: The total number and distribution of alleles and genotypes in a population; a population with a very high genetic diversity would have many alleles and genotypes, all evenly distributed or with approximately equal frequency.

genetic drift: Change in gene frequencies in a population owing to chance.

genetic modification: Alteration of a organism's genetic material by manipulation in the laboratory. The addition of new genes or the removal of genes are examples.

genome: The complete amount of DNA found in the nucleus of a normal cell, expressed as a particular number of chromosomes; for example, a human cell has a genome of forty-six chromosomes.

genotype: The complete genetic makeup of an organism, regardless of whether these genes are expressed. *See also* phenotype.

genotype frequency: The relative abundance of a genotype in a population; to calculate, count the number of individuals with a given genotype in the population and divide by the total number of individuals in the population.

genus (*pl.* genera): A group of closely related species having many traits in common and descended from a common ancestor; for example, *Felis* is the genus of cats, and it includes the species *Felis catus* (the domestic cat) and *Felis couguar* (the cougar or mountain lion).

geochemical cycles: *See* biogeochemical cycles.

global ecology: The study of the impacts of such factors as global warming, pollution, and disease on organisms and ecosystems worldwide. Much of global ecology considers the ecological impacts of human-driven influences such as international travel, trade, the built environment, and the use of petrochemicals.

global extinction: The loss of all members of a species; that is, extinction whereby all populations of a species disappear or are eliminated. *See also* local extinction.

global warming: The theory that the atmosphere is becoming warmer over time as a result of an increase in greenhouse gases, resulting in climate change, melting ice caps, rising sea levels, severe weather events, and their impacts on the world's living organisms.

glycosides: Compounds produced by plants that combine a sugar, usually glucose, with an active component. Potentially poisonous glycosides include the cyanogenic, cardioactive, anthraquinone, coumarin, and saponin glycosides.

gradient analysis: A method for studying the distribution of species along an environmental gradient.

gradualism: A model of evolution in which transformation from ancestor to descendant species is a slow, gradual process spanning millions of years.

grassland: A biome in which the dominant plants are grasses.

grazer: An organism that feeds primarily on grasses.

Green Revolution: Several decades of dramatic advances in yield and quality of crop species. This was the outcome of attempts to increase food production begun in the 1940's.

greenhouse effect: The process whereby the infrared radiation from the earth is absorbed by the atmosphere, keeping it from escaping into space. Because it was once believed that the glass panes of greenhouses acted similarly, this phenomenon was termed the "greenhouse" effect, although it has subsequently been shown that greenhouses work differently (by trapping heated air and not allowing it to blow away).

greenhouse gases: gases, including carbon dioxide, water vapor, and methane, store heat more efficiently than others and contribute to the magnification of the so-called greenhouse effect.

gregarious: Forming groups temporarily or permanently.

gross primary productivity: The amount of the sun's energy actually assimilated by autotrophs. *See also* net primary productivity; secondary productivity.

habitat: The physical environment, usually that of soil and vegetation as well as space, in which an animal lives.

habitat fragmentation: Conversion of a natural, contiguous landscape into smaller patches usually due to human activity. Grids of roads, gas lines, cluster housing, and power lines all contribute to habitat fragmentation.

habitat selection: Process of choosing a home range, territory, nesting site, or feeding site on the basis of specific features of the habitat that an organism is best adapted to exploit.

habituation: The process of learning to ignore irrelevant stimuli that previously produced a reaction.

haploid: Having one set of chromosomes and one of each kind of gene. Gametes (eggs or sperm) are usually haploid. *See also* diploid.

Hardy-Weinberg law: A concept in population genetics stating that, given an infinitely large population that experiences random mating without mutation or any other such affecting factor, the frequency of particular alleles will reach a state of equilibrium, after which their frequency will not change from one generation to the next.

hazardous waste: Any waste product that is toxic to living beings or threatens life.

hemiparasite: A plant (such as mistletoe) with some chlorophyll which lives as a partial parasite.

herbicide: Chemical that is lethal to plants.

herbivore: An animal that eats living plants, usually specialized to digest cellulose.

heritability: The extent to which variation in some trait among individuals in a population is a result of genetic differences.

heterochrony: Any phenomenon in which there is a difference between the ancestral and descendant rate or timing of development.

heteroecious: Describes fungi that spend parts of their life cycle on entirely different species of plants. Black stem rust of wheat spends part of its life cycle on wheat and part on American barberry.

heterotrophs: Organisms that cannot make their own food from simpler materials but must ingest and digest complex molecules made by other organisms. Animals are heterotrophs, but so are nongreen plants such as Indian pipes. *See also* consumer.

heterozygous: Having two different alleles at a particular gene locus.

hibernation: A sustained period of torpor (lack of activity) triggered by cold environmental conditions, achieved when an animal reduces its metabolic rate. *See also* estivation.

hierarchy: A social structure in which animals are dominated by those higher on the linear ladder.

home range: Geographic area used by an individual, pair, or group for their daily, seasonal, and sometimes their yearly activities; the defended portion of the home range is called a territory.

homeostasis: The dynamic balance between body functions, needs, and environmental factors which results in internal constancy.

homologous: Referring to chromosomes that are identical in terms of types of genes present and the location of the centromere; because of their high degree of similarity, homologous chromosomes can synapse and recombine during prophase I of meiosis.

homozygote: A diploid organism that has two identical alleles for a particular trait; a person with blood type A would be homozygous if he had two A alleles.

horizons (soil): Layers of soil with different appearance and different chemical composition. In a forest, the horizon closest to the top will contain more organic matter than will the deeper layers.

hormone: Chemical compound produced in small amounts in one part of an organism that has an effect in a different part of the same organism. Auxins, gibberellins, cytokinins, ethylene, estrogen, progesterone, oxytocin, and various pheromones are examples.

host: In a parasitic relationship, the organism that is giving up energy-containing molecules to the parasite.

hybrid: An organism resulting from the crossing of two species.

hybrid vigor: The tendency of hybrids to be larger and more durable than their parent species; also called heterosis.

hybrid zone: An area with a population of a species composed of individuals with characteristics of one or more species that have interbred.

hybridization: A process of base-pairing involving two single-stranded nucleic acid molecules with complementary sequences; the extent to which two unrelated nucleic acid molecules will hybridize is often used as a way to determine the amount of similarity between the sequences of the two molecules; hybridization is fairly common among wind-pollinated plants, while hybridization is quite uncommon among higher animals.

hydrologic cycle: Earth's cycle of evaporation and condensation of water, which produces rain and maintains oceans, rivers, and lakes.

hydrophilic: Loving water, capable of mixing with or growing in water.

hydrosphere: The waters of the earth and the regions where they are found, including freshwater lakes and rivers, the oceans, inland seas, and then frozen water of the polar zones.

immigration: The movement of organisms into an area; a one-way movement into a habitat type.

imprinting: A specialized form of learning characterized by a sensitive period in which an association with an object is formed.

inbreeding: Mating between relatives, an extreme form of positive assortative mating. *See also* outbreeding.

inbreeding depression: Weakening, or lowering of the fitness, of offspring as a result of inbreeding, caused when alleles that decrease fitness drift to fixation.

indicator species: Any species that is among the first to be degraded when an ecosystem is compromised, indicating the ecological impact of environmental change.

individual ecology: *See* behavioral ecology; physiological ecology.

industrial melanism: The rapid rise in frequency of the melanic form in many moth species downwind of manufacturing sites, associated with the advent of industrial pollution.

infauna: Animals that live in the sea floor.

inflorescence: The group of flowers that forms at the top of a flower stalk. Inflorescences may be compact, like that of a daisy, or loose, like that of a mustard.

innate: Inborn, possessed from birth, or determined and controlled largely by the genes.

insectivore: Any of an order of small, nocturnal mammals, including shrews, moles, and hedgehogs, *Insectivora*, or generally any animal that feeds on insects.

insectivorous plant: Plant that traps insects and digests them for the nitrogen-containing compounds they possess. Plants of this type grow in boggy soil that is typically low in nitrogen.

instinct: Any behavior that is completely functional the first time it is performed.

integrated pest management (IPM): The practice of integrating insect, animal, or plant management tactics, such as chemical control, cultural control, biological control, and plant resistance, to maintain pest populations below damaging levels in the most economical and environmentally responsible manner.

interbreeding: The mating of closely related individuals which tends to increase the appearance of recessive genes.

interference: The act of impeding others from using some limited resource.

interfertile: Able to breed and produce fertile offspring.

intermediate host: An animal species in which nonsexual developmental stages of some commensals and parasites occur.

interspecific competition: Competition between species that need the same limited resource.

intertidal marine zone: The shallow marine biome that lies at the land's edge.

intraspecific competition: Competition between members of the same species for a limited resource.

intrinsic rate of increase: The growth rate of a population under ideal conditions, expressed on a per individual basis.

introgression: The assimilation of the genes of one species into the gene pool of another by successful hybridization.

invasive species: Nonnative (also termed "exotic") organisms that can outcompete native species. *See also* exotic species.

iridescent: Showing the colors of the rainbow depending light reflection.

irruption: A sudden increase in the size of a population, usually attributed to a particularly favorable set of environmental conditions.

island biogeography: a distinct subdiscipline of biogeography that considers biodiversity and island size, ecological heterogeneity, proximity to continents, isolation and endemism, island size and location and rates of immigration, colonization, and extinction. Island biogeographers frequently have debates arguing the relevance of dispersal versus vicariance.

isolation: *See* reproductive isolation.

isotherm: A line or boundary imagined on the earth's surface that connects points having the same temperature at a given time.

K strategy [selection]: A reproductive strategy typified by low reproductive output; common in species living in areas having limited critical resources.

keystone species: A species that determines the structure of a community, usually by predation on the dominant competitor in the community.

kin selection: A phenomenon by which acts of altruism can help pass on genes for altruism by improving the survival of kin and their offspring.

landscape ecology: A relatively new field of ecology, the study of how ecosystems, including the built environment, are arranged and how their arrangements affect the wildlife and environmental conditions that form them. Landscape ecologists examine land patterns (topography, water, forest cover, and human uses) and how these affect wildlife populations.

lichen: An organism formed by the symbiotic relationship between a fungus and an alga. The alga is protected from drying by the fungus, while the fungus receives food molecules from the alga.

life cycle: The sequence of development beginning with a certain event in a organism's life (such as the fertilization of a gamete), and ending with the same event in the next generation.

life expectancy: The probable length of life remaining to an organism based upon the average life span of the population to which it belongs.

life span: The maximum time between birth and death for the members of a species as a whole.

life table: A chart that summarizes the survivorship and reproduction of a cohort throughout its life span.

limnology: The study of the physical, chemical, climatological, biological, and ecological aspects of lakes.

litter: The offspring produced in a single birth; also referred to as a clutch.

littoral zone: The shore zone of a lake, where macrovegetation grows.

local extinction: The loss of one or more populations of a species, but with at least one population of the species remaining. *See also* global extinction.

logistic growth: A pattern of population growth that involves a rapid increase in numbers when the density is low but slows as the density approaches the carrying capacity.

macroevolution: Large-scale evolutionary processes that result in major changes in organisms and allow them to occupy new adaptive niches or develop novel body plans.

marine ecology: A branch of both ecology and oceanography that investigates the ocean zones and the interactions of their biotic and abiotic components, including all components impinging on the oceans such as seabirds and coastal zones.

mark-capture-release methods: Methods of studying populations by capturing and marking some members, then releasing them into the wild and periodically recapturing them, or capturing successive samples, in order to track their status at different points in time.

mass extinction: An event in which a large number of organisms in many different taxa are eliminated; there have been five such events in the history of life that resulted in the disappearance of more than 75 percent of all species.

mate competition: Competition among members of one sex for mating opportunities with members of the opposite sex.

maximum sustainable yield (MSY): The rate of harvest of natural resources such as fisheries and timber that can be maintained indefinitely through active human management of those resources.

mediterranean scrub: Type of vegetation found in certain places, such as Southern California, which experience wet winters and long, dry summers.

metabolites: Compounds formed as the result of biochemical pathways in an organism.

metamorphosis: An abrupt change from one life-form to another, such as from a larval body form, accompanied by many physiological changes in the determination, differentiation, and distribution of cells, into an adult body form.

microbivores: Organisms that eat microbes.

microevolution: Small-scale evolutionary processes resulting from gradual substitution of genes and resulting in very subtle changes in organisms.

microphages: Animals that feed on small microscopic particles suspended in water or deposited on bottom sediments.

migration: The movement of individuals resulting in gene flow, changing the proportions of genotypes in a population.

mimicry: The resemblance of one species (the model) by one or more other species (mimics), such that a predator cannot distinguish among them.

molecular clock: Accumulation of genetic changes that develops when two species diverge. The longer two species have been separate, the more changes will be evident. The clock does not tick at the same rate in every species or every molecule.

molecular ecology: The study of natural and introduced microbial populations and their environments and ecological implications of the release of recombinant organisms. The development of molecular genetic techniques has led to this relatively new field for addressing ecological questions and issues.

monoclimax theory: The theory, promulgated by Frederic A. Clements, that all communities within a given climatic region, despite initial differences, eventually develop into the same climax community. *See also* polyclimax theory.

monocropping: The common agricultural practice of planting and growing single crops, usually on a large tract of land, requiring the use of chemical fertilizers and pest controls.

monoculture: A single crop used in monocropping.

monohybrid: An organism that is hybrid with respect to a single gene.

monsoon forest: Forest type found in tropical regions of the world where there are annual periods of high rainfall.

mortality rate: The number of organisms in a population that die during a given time interval.

Müllerian mimicry: Mimicry in which the mimic is toxic.

multiple use: Resource use in which land supports several concurrent managed uses rather than single uses over time and space.

mutation: A change in the genetic sequence of an organism, sometimes leading to an altered phenotype.

mutualism: A type of commensalism or symbiosis in which both symbiotes benefit from the association in terms of food, shelter, or protection.

mycobiont: The fungal part of a lichen.

mycorrhiza (*pl.* mycorrhizae): A symbiotic relationship between a root and a fungus in which the fungus lives either in or on the root and gains food from it. The fungus increases absorption of water and minerals for the root.

N: A standard abbreviation for the size of an actual population; if ñ, it is an estimated value.

natality rate: Birthrate: the number of individuals that are born into a population during a given time interval.

natural selection: The process of differential survival and reproduction that leads to heritable characteristics that are best suited for a particular environment.

necroparasite: A parasite that consumes or lives in dead tissue.

nekton: An aquatic organism that has the ability to swim.

neoteny: Either the retention of immature characteristics in the adult form or the sexual maturation of larval stages; it results in new kinds of adult body plans.

neritic zone: The marine zone that begins at a depth of about 600 feet (180 meters), where the gradual slant of the continental shelf becomes a sharp tilt toward the ocean floor, and extends to the ocean bottom.

nest parasite: Also called brood parasite; an individual (or species) that lays its eggs in the nest of another individual (or species) and does no parenting at all.

net primary productivity: The amount of energy, after plant respiration, that is potentially available to primary consumers. *See also* gross primary productivity; secondary productivity.

neutral mutation: A mutation with no observable effect on the phenotype of the cell or organism in which it occurs.

neutralism: A mutualism in which the two species are neither helped nor harmed.

niche: An organism's role in its habitat environment, such as food producer, decomposer, parasite, plant eater (herbivore), meat-eater (carnivore). The sum of environmental conditions necessary for the survival of a population of any species, including food, shelter, habitat, and all other essential resources.

nitrogen cycle: The biogeochemical cycle of the element nitrogen.

nitrogen fixation: Process by which bacteria convert atmospheric nitrogen to nitrogen-containing compounds, such as amino acids.

nocturnal: Active at night.

nomadic: Moving about from place to place according to the state of the habitat and food supply.

nonrandom mating: Mating that occurs whenever every individual does not have an equal chance of mating with any other member of the population.

nonrenewable resource: A natural resource, such as copper, coal, aluminum, and oil, that exists in a fixed amount and cannot be naturally replenished at the rate it is being mined or removed.

nonruminating: Digesting grasses without chewing cud.

numerical response: The abundance of predators dependent upon the abundance of prey in a habitat.

nutrient cycles, nutrient cycling: Large-scale movements of elements (such as carbon, nitrogen, and water) through the living and nonliving portions of an ecosystem.

old-growth forest: Ancient forests in which many trees are hundreds of years old and which are among the richest ecosystems on earth. Considered nonrenewable.

omnivore: An animal that eats both plant material and animal material.

ontogeny: The successive stages during the development of an animal, primarily embryonic but also postnatal.

opportunistic species: An invasive species that drives out native species by competing with them for resources.

ordination: A method for collapsing community data for many species in many communities along several environmental gradients onto a single graph that summarizes their relationships and patterns.

organic: Living. At the molecular level, containing carbon atoms as a primary component.

organism: Any individual, self-contained life form, whether unicellular or multicellular, whether protist, fungal, algal, plant, or animal.

orientation: An inherent sense of geographical location or place in time.

outbreeding: Interbreeding of stocks of a species that are unrelated to each other. *See also* inbreeding.

outgroup: A group of organisms only distantly related to the groups being examined in a cladistic study.

overfishing: Harvesting so many fish, including sexually immature fish, that fishing becomes commercially inviable and ultimately leading to the species' extinction.

overgrazing: Destruction of vegetation by allowing too many grazing animals to graze beyond the carrying capacity of an area of rangeland.

overhunting: Harvesting so many terrestrial species (usually mammals or reptiles), including sexually immature individuals, that the species is ultimately threatened with extinction.

ozone layer: The ozone-enriched layer of the stratosphere that filters out some of the sun's ultraviolet radiation, which causes skin and other types of cancer.

P generation: Parental generation; the original individuals mated in a genetic cross.

pair-bonding: Prolonged and repeated mutual courtship display by a monogamous pair, serving to cement the pair bond and to synchronize reproductive hormones.

paleoecology: The study of past ecosystems and environments.

parasite: Any organism that lives on or in another living organism, the host, and obtains its food from that host.

parasite-mix: All the individuals and species of symbiotes living concurrently in a host.

parasitism: The state and activities of being a parasite.

parasitoid: An insect, especially a wasp, that spends its larval stage in the body of another insect and eats it, eventually emerging as a free-living adult.

patch: A habitat fragment within a landscape, often occuring in a patchwork of artificial and natural habitats strewn across a landscape in haphazard and unplanned fashion.

pecking order: A dominance hierarchy in which the top individual can threaten and force into submission any individual below it, the number two individual can threaten anyone except number one, and the lowest-ranked individual can threaten no one and must submit to everyone.

pelagic: The area of open water in the oceans; organisms that occur in the water column.

per capita rate of increase: The difference between per capita births and per capita deaths, defined as *r*. Also called the per capita rate of growth.

periodicity hypothesis: The proposal that mass extinctions have occurred approximately every 26 million years over the past 250 million years.

permafrost: Permanently frozen layer of soil underlying tundra vegetation.

perturbation: Factors such as diseases, parasites, fire, and deforestation, that disrupt ecosystems from within.

pesticide: Chemical or biological substances designed to kill unwanted plants, fungi, or animals that interfere, directly or indirectly, with human activities. *See also* biopesticides.

phenology: The science that examines the relationships between climates and climatic conditions and animal behaviors such as migrations or parts of the life cycle such as flowering in plants.

phenotype: The visible or outward expression of the genetic makeup of an individual. *See also* genotype.

pheromone: A chemical produced by one member of a species that influences the behavior or physiology of another member of the same species.

phosphorus cycle: The biogeochemical movement of phosphorus through an ecosystem.

photoperiodism: Regulation of a plant process, such as flowering, by the relative length of light and dark periods in a twenty-four-hour day.

photorespiration: The process that occurs during photosynthesis when the ratio of carbon dioxide to oxygen becomes too low. It results in the production of less sugar than during regular photosynthesis, and some carbon dioxide is formed.

photosynthesis: The process by which green plants and algae, called primary producers, use sunlight as energy to convert carbon dioxide and water into energy-rich compounds such as glucose.

phyletic: related by or having a common ancestral line in evolutionary terms.

phylogenetics: The study of the developmental history of groups of animals.

phylogeny: The evolutionary history of taxa, such as species or groups of species; order of descent and the relationships among the groups are depicted.

physiological ecology: Sometimes called "autoecology," "ecophysiology," or "comparative physiology," a type of individual ecology that examines how life-forms function mechanically and physiologically in their environments and how such factors as temperature, seasons, soil, and nutrients affect survival and reproduction of those organisms. Analyzes organismic adaptations from an engineering perspective in an evolutionary context to determine the relationship between individuals' performance attributes, populations, and communities.

physiology: The study of the functions, activities, and processes of living organisms.

phytophagous: Animals, also referred to as herbivorous, that feed on plants.

phytoplankton: Small plants, often single-celled, that float in water. Phytoplankton in the ocean are responsible for much of the earth's oxygen production.

pioneer species: The earliest, hardy organisms that begin colonizing an area in the first stage of ecological succession.

plankton: Small plants and animals that float freely in water; their small size prevents them from having to swim to keep from sinking.

plant ecology: The study of plant life (often including fungal and algal forms) on all levels of ecology, from physiological to population to community to ecosystem ecology. Applications are particularly important in agriculture.

poikilotherm: Cold-blooded or ectothermic; any organism having a body temperature that varies with its surroundings; in general, reptiles, amphibians, fish, and invertebrates.

pollination: Transfer of pollen to a stigma in angiosperms or an ovule in gymnosperms.

pollution ecology: The study of the impacts of water, air, and waste pollution on populations, communities, and ecosystems.

polyclimax theory: Within a given climatic region, there can be many climaxes. *See also* monoclimax theory.

polygamy: A mating system in which one male mates with several females (polygyny) or one female mates with several males (polyandry).

polygenic inheritance: Expression of a trait depending on the cumulative effect of multiple genes; human traits such as skin color, obesity, and intelligence are thought to be examples of polygenic inheritance.

polymorphism: The occurrence of two or more structurally or behaviorally different individuals within a species.

polyphyletic: Having similar characteristics but originating from more than one ancestor.

polyploidy: Genetic condition in which the hereditary material is present in three or more complete sets. Plants may be triploid (three sets of chromosomes), tetraploid, or have even greater numbers.

population: A group of individuals of the same species that live in the same location at the same time.

population analysis: The study of factors that influence growth of biological populations.

population density: The number of individuals in a population per unit area or volume.

population distribution: Variations in the density of a population in a particular area.

population dynamics: The patterns of population growth and decline over the population's existence, influenced by reproductive rates, predator-prey relationships, carrying capacity, and other such factors.

population ecology: The study of the growth and decline of groups of individuals of the same species, and how these fluctuations function in relation to other populations in the same ecosystem. Examines such factors as the availability of food and hence predation, herbivory, and mutualisms. Community ecology is closely related to population ecology but focuses on the interactions between populations of different species.

population fluctuations: Changes in population size over time.

population genetics: Branch of genetics that examines the movement of genes through populations. A population geneticist might study genetic drift, founder effect, or Hardy-Weinberg theorem.

population regulation: Stabilization of population size by factors such as predation and competition, the relative impact of which depends on abundance of the population in a habitat.

positive feedback loop: Situation in which a change in a certain direction provides information that causes a system to change further in the same direction. This can lead to a runaway or vicious cycle.

predation: The act of killing and consuming another organism. The organism that gains its nutrition in this way is the predator; the organism that serves as the source of nutrition is the prey.

primary consumer: An organism that get its nourishment from eating primary producers, which are mostly green plants and algae.

primary metabolites: Sugar phosphates, amino acids, lipids, proteins, and nucleic acids, all of which comprise the basic molecules necessary for a cell to function.

primary producers: In an ecosystem, those organisms that form foods from energy and simple chemical molecules.

primary succession: Succession in which the initial seral stage, or pioneer community, begins on a substrate devoid of life or unaltered by living organisms. *See also* secondary succession.

producers: *See* primary producers.

productivity: The rate of accumulation of biomass. *See also* gross primary productivity; net primary productivity; secondary productivity.

promiscuity: A mating system in which sexual partners do not form lasting pair bonds; their relationship does not persist beyond the time needed for copulation and its preliminaries.

protective mimicry: Use of both color and form to mimic an inanimate feature of the environment.

punctuated equilibrium: The idea that new species form during relatively short speciation events (a few generations) and then persist for millions of years in equilibrium (relatively unchanged) until they go extinct or speciate again.

quadrat: A sample plot of a specific size and shape used in one method of determining population size or species diversity.

r strategy (r selection): A reproductive strategy involving high reproductive output; found often in unstable or previously unoccupied areas.

rain forest: Biome found in regions of the world where rainfall is high and there are no dry periods. Rain forests occur in the tropics and along the northwest coast of the United States.

random genetic drift: The random change of gene frequencies because of chance, especially in small populations.

random mating: The assumption that any two individuals in a population are equally likely to mate, independent of the genotype of either individual; this is equivalent to saying that all the gametes of all the individuals in a population are placed into a large pool, from which gametes are paired at random.

rangeland: Open land of a wide variety of types, including grasslands, shrublands, marshes, and meadows as well as some desert and alpine land.

recapture: *See* mark-capture-release methods.

reciprocal cross: A mating that is the reverse of another with respect to the sex of the organisms that possess certain traits; for example, if a particular cross were tall male X short female, then the reciprocal cross would be short male X tall female.

reciprocal relationship: Any type of coevolved, highly interdependent relationship between two or more species.

reciprocal sacrifice: One explanation for acts of altruism among unrelated animals; an individual sacrifice is made under the assumption that a similar sacrifice may in turn aid the individual in the future.

recombinant DNA: DNA molecules that are the products of artificial recombination between DNA molecules from two different sources; important as a foundation of genetic engineering.

recombination: An exchange of genetic material, usually between two homologous chromosomes; provides one of the foundations for the genetic reassortment observed during sexual reproduction.

red tide: An algal bloom that can be red, orange, brown, bright-green, or even blooms that do not discolor the water in which they grow, usually caused by dinoflagellates and often toxic.

reef: A carbonate structure that possess an internal framework that traps sediment and provides resistance to wave action, thus forming habitat for a rich and diverse marine community. Types of reefs include atoll, barrier, fringing, and patch reefs.

reforestation: The growth of new trees in an area that has been cleared for human activities, either occurring naturally or initiated by people.

relict population: A remnant population of otherwise extinct organisms.

replication: Copying of a DNA molecule, so that the two molecules formed are identical to each other and to the original molecule. The two molecules formed are each composed of one strand from the original DNA molecule and one newly synthesized strand.

reproductive (genetic) isolation: Describes any of many mechanisms that prevent one organism from sexually reproducing with a second.

reproductive strategy: A set of traits that characterizes the successful reproductive habits of a group of organisms.

reservoir host: A host species other than the one of primary interest in a given research study.

resilience stability: Stability exhibited by a community that changes its structure when disturbed but returns to its original structure when the disturbance ends.

resistance: The ability of an organism, population, or community to survive in the face of natural or human-made threats.

resource: A requirement for life, such as space for living, food (for animals), or light (for plants), not including conditions such as temperature or salinity.

resource-holding potential: The ability of an individual to control a needed resource relative to other members of the same species.

respiration: The utilization of oxygen; in air-breathing vertebrates, the inhalation of oxygen and the exhalation of carbon dioxide.

restoration ecology: Restoration ecology is the study and implementation of ways to return degraded or deteriorating communities and ecosystems to their original condition. Restoration ecologists work to restore habitat and return endangered species to viable numbers; they do not seek to restore extinct species or re-create ancient habitats. *See also* conservation biology.

riparian ecosystem: An ecosystem in and around a river.

ritualization: An evolutionary process that formalizes the context and performance of a display so that its meaning is clear and straightforward.

runoff: Surface water effluent from precipitation, irrigation, or other human activity and the chemicals and other materials it ultimately carries into aquatic systems.

salinization: The accumulation of salts in soil.

saprophyte: Type of fungus or plant that gets its nutrition by digesting dead plants. Many mushrooms grow as saprophytes.

saprovore: An organism that consumes dead or decaying plant or animal matter.

savanna: Biome type characterized by widely spaced trees separated by open, grassy regions.

scale of being: An arrangement of life-forms in a single linear sequence from lower to higher; also called a chain of being.

scavenger: An animal that feeds on the carcasses of other animals.

scrub community: Plant community characterized by stunted trees and shrubs that may be widely spaced. Typical of poor soils.

secondary consumers: Carnivorous animals that eat herbivorous animals (primary consumers).

secondary metabolite: A biochemical that is not involved in basic metabolism, often of unique chemical structure and capable of serving a defensive role for an organism.

secondary productivity: The rate at which animals produce their organic matter by feeding on other organisms. *See also* gross primary productivity; net primary productivity.

secondary succession: Succession that starts in areas where an established community has been disturbed or destroyed by natural forces or by human activities. *See also* primary succession.

seed dispersal: Process by which the seeds of a plant are distributed over a wide area, away from the parent plant. Many seeds are adapted for dispersal by wind or animals.

selection: A process that prevents some individuals from surviving and propagating while allowing others to do so. *See also* natural selection.

selective pressure: Evolutionary factors that favor or disfavor the genetic inheritance of various characteristics of a species.

self-pollination: When pollen from the anther of a flower lands on the stigma of the same flower.

semelparous species: Species, such as the Chinook salmon, that reproduce only once before dying.

semiochemical: A chemical messenger that carries information between individual organisms of the same species or of different species; pheromones and allelochemics are semiochemicals, but hormones are not.

sensitization: An arousal or an alerting reaction which increases the likelihood that an organism will react; also, a synonym for loss of habituation with increased intensity of response.

sequester: To store a material derived from elsewhere. In defenses, some predators sequester defensive properties from their prey to defend themselves from their own predators.

sex-role reversal: Generally used to refer to species in which the male does most of the parenting.

sexual dimorphism: A difference in structure or behavior between males and females.

sexual reproduction: Reproduction of cells or organisms involving the transfer and reassortment of genetic information, resulting in offspring that can be phenotypically and genotypically distinct from either of the parents.

sexual selection: The process that occurs when inherited physical or behavioral differences among individuals cause some individuals to obtain more matings than others.

signal: Information transmitted through sound, such as bird calls, or through sight, such as body posture.

signal pheromone: Nearly synonymous with releaser pheromone, but used with mammals to remove the suggestion of a programmed response and to indicate a more complex response.

sit-and-wait foraging: Sitting in one place, waiting, and attacking prey as they move.

slash-and-burn (swidden) agriculture: An agricultural practice in which forestland is cleared and burned for use in crop and livestock production.

SLOSS: An acronym for single large or several small preserves. This mirrors and reflects the current discussion as to how best preserve, protect, and manage wildlife by use of one large park or preserve or several smaller parcels.

social ecology: A branch of ecology, related to sociology, that critiques the relationship between environmental destruction and social structure or political ideology.

social grooming: An activity maintaining social interaction, whereby debris is removed from a primate's hair.

sociality: The tendency to form and maintain stable groups.

sociobiology: The study of the biological basis of the social behavior of animals.

soil ecology: The study of soil as an ecosystem, including the interactions of both abiotic and biotic components of soil: bacterial, fungal, plant (humus, living roots), and animal (from protozoa and nematodes to insects and earthworms). Soil ecology extends beyond the physical borders of soil to include the impact of soil on aboveground life-forms such as larger plants and animals, as well as processes (geochemical cycles, erosion, human agricultural practices) that impact soil. *See also* agricultural ecology.

soil erosion: Loss of topsoil because of erosion due to runoff, winds, and other forces.

soldiers: In insect societies, large workers that defend the colony and often raid other colonies.

somatic hybrids: Plants that result from the fusion of two somatic (nonreproductive) cells.

specialists: Species with narrow niches that are not well generalized, often able to live in only one habitat.

speciation: The evolution of new species as a result of geographic, physiological, anatomical, or behavioral factors that prevent previously interbreeding natural populations from breeding with each other any longer.

species: A group of animals capable of interbreeding under normal natural conditions; the smallest major taxonomic category.

species diversity: The variety of different organisms at the species taxonomic level, a value that combines measures of both species richness and species evenness.

species loss: Extinction.

species selection: The idea that species are independent entities with their own properties, such as birth (speciation) and death (extinction); a higher level of selection above that of natural selection is postulated to take place on the species level.

species-specific: Innate and exclusive to a single species.

status badge: A visual feature that, based on its size or color or some other variation, indicates the social status of the bearer.

stereotyped behavior: An unlearned and unchanging behavior pattern that is unique to a species.

stimulus: Any environmental cue that is detected by a sensory receptor and can potentially modify an animal's behavior.

strategy: A behavioral action that exists because natural selection favored it in the past (rather than because an individual has consciously decided to do it).

stratification: Division of a community into layers, such as canopy, shrub, and herb layers of a forest.

stress: A pressure on an organism, population, or community as a result of ecosystem disturbance or an inadequate resource such as water or nutrients.

stromatolite: Fossilized masses of cyanobacteria that represent some of the first evidence of photosynthetic life on earth.

subspecies: A group or groups of interbreeding organisms within a single species that are distinct and separated from similar related groups but not reproductively isolated.

substrate: The substance, such as soil or bark, on which a plant grows. Also, in biochemical reactions, the chemical compound upon which an enzyme acts.

succession: Change in a plant or animal community over time, with one kind of organism or plant being replaced by other organisms or plants in a more or less predictable pattern. *See also* primary succession; secondary succession.

superorganism concept: An "organism" composed of more than one member, such as an insect society, reflecting the remarkable degree of coordination between individuals.

survivorship: The pattern of survival exhibited by a cohort throughout its life span.

sustainable development: The study and implementation of methods of agriculture, mining, timber harvesting, and other human interactions with the natural environment in a way that ensures minimal or no impact on the original ecosystems, to the end of conserving natural resources and minimizing impact on the organisms that occupy the habitats involved. The term "sustainable" is often coupled with more specific disciplines: "sustainable agriculture," "sustainable forestry," and so on. *See also* restoration ecology.

swidden agriculture: *See* slash-and-burn agriculture.

switching: The phenomenon that occurs when a predator has a choice of several prey species and learns to prefer one of them over a previous prey; if the preferred prey is sufficiently abundant, the predator will begin to concentrate on the preferred prey and may change its searching and other prey-related behaviors.

symbiont: A member of a symbiotic relationship.

symbiosis: A type of coevolved relationship between two species in which both participants benefit; a type of mutualism.

sympatric: Living in the same place, not separated by a barrier that would prevent interbreeding.

sympatric speciation: Evolution of two groups into separate species while living in overlapping geographic locations.

synergism: The result of the interactions of a number of agents, operations, organisms, and other factors such that the final effect is greater than the sum of the individual effects.

systematics: The subdivision of biology that deals with the identification, naming, and classification of organisms and with understanding the evolutionary relationships among them. *See also* classification; taxonomy.

taiga: Biome located across Canada, northern Europe, and northern Asia that consists of forests of spruces, firs, and birches. *See also* boreal forest.

taxonomy: A classification scheme for organisms based primarily on structural similarities; taxonomic groups consist of genetically related animals. *See also* classification; systematics.

temperate deciduous forest: Biome located in eastern North America south of the taiga, central Europe, and eastern China, characterized by forests of tree species, most of which lose their leaves every autumn.

temperate mixed forest: A subdivision of the temperate deciduous forest in which pines share importance with deciduous trees. These forests are usually found on sandy soils in the southeastern United States.

temporal variation: Variation across time.

terrestrial: Living on land.

territorial behavior: The combination of methods and actions through which an animal or group of animals protects its territory from invasion by other species.

threat display: A territorial behavior exhibited by animals during defense of a territory, such as charging, showing bright colors, and exaggerating body size.

threatened species: Animals or plants, as designated by official bodies such as the U.S. Fish and Wildlife Service, whose members are so few in number that they may soon become endangered and then extinct.

topography: The structure and configuration of the earth's surface, including its natural and human-made features.

toxin: Any substance, such as the venom in snakes or spiders, that is toxic to an animal.

trace element: Chemical element required in very small quantities for nutrition.

transgenic: Possessing one or more genes from another organism, a concept important for the study of genetic mutations.

transpiration: Evaporation of water from the leaves and stems of plants. Most transpiration occurs through open stomata.

trophic levels: Positions at which specific organisms obtain their nutrition within a food chain. Plants and photosynthetic algae, which are primary producers, occupy the first trophic level, followed by primary and secondary consumers, scavengers, and decomposers.

tropical rain forest: Biome found in tropical regions throughout the world that experience rainfall year-round. These are the most biodiverse ecosystems, often containing thousands of species of plants in a single acre.

tropism: Plant growth response to an external stimulus. The plant may grow toward or away from the stimulus. Includes geotropism, gravitropism, heliotropism, phototropism, and thigmotropism.

tundra: Regions where no trees grow because of frozen soil or extreme water runoff due to steep grades (at high altitudes) are known as tundra. *See also* alpine tundra; Arctic tundra.

understory: In a rain forest, the shorter trees and shrubs that grow under and live in the shade of the canopy formed by the crowns of the tallest trees.

uniformitarianism: The belief that the earth and its features are the result of gradual biological and geological processes similar to the processes that exist today.

urban ecology: The study of the ecology of cities and human settlements, including energy and water flows, resource use, and how and where humans build. *See also* landscape ecology.

urban heat island: A relatively hot area in the atmosphere above an urban area produced by a concentration of cars, factories, reflective surfaces, and other heat-producing factors.

urbanization: The process of converting natural habitats into urban complexes.

vector: An organism, such as a baterium, a mosquito, a virus, or a rodent or other mammal, that transmits pathogens from one host to another.

veld: A grassland ecosystem, especially in southern Africa.

venom: A toxic substance that must be injected (instead of ingested) to immobilize or kill prey.

virion: A virus particle consisting of genetic material surrounded by a protein coat.

viroid: Small, viruslike infectious molecules of RNA without protein coats.

virus: A microscopic infectious particle composed primarily of protein and nucleic acid; bacterial viruses, or bacteriophages, have been important tools of study in the history of molecular genetics.

visual predation: Catching prey (such as insects) by sighting them visually, judging their exact position and distance, and pouncing on them.

warm-blooded: Referring to animals whose body temperatures are maintained at a constant level by their own metabolisms.

warning coloration: The bright colors seen on many dangerous and unpalatable organisms that warn predators to stay away.

weeds: hardy plants, usually invasive species, that successfully compete with native plants to the latter's eventual weakening and extinction.

wetlands: Transitional areas between aquatic and terrestrial habitats, home to a variety of flood-tolerant and salt-tolerant species.

wildlife: All living organisms; traditionally, the term included only mammals and birds that were hunted or considered economically important.

workers: Sterile, wingless female social insects.

zonation: The distribution of organisms and other elements of ecosystems into regular or distinct biogeographical zones.

zoogeography: The study of the distribution of animals over the earth.

zoology: The study of the classification, anatomy, and physiology of animal life.

zooplankton: Small animals, often single-celled, that float or swim weakly.

zygote: A diploid cell produced by the union of a male gamete (sperm) with a female gamete (egg); through successive cell divisions, the zygote will eventually give rise to the adult form of the organism.

WEB SITES

Biodiversity and Ecosystem Function Online
http://www.abdn.ac.uk/ecosystem/bioecofunc
An "informal collaboration between several research institutes within the United Kingdom that share similar research interests. Hosted by the University of Aberdeen, Scotland, the site aims to be a resource for researchers specifically concerned with how the richness of species (or functional groups) affects ecosystem function. A list of the participants and a link to their respective research interests can be accessed through the 'Project Partners' page." Offers a bibliography on ecosystem ecology.

British Ecological Society
http://www.britishecologicalsociety.org
The BES "is an active and thriving organisation with something to offer anyone with an interest in ecology. Academic journals, teaching resources, meetings for scientists and policy makers, career advice and grants for ecologists" are among its activities. The Web site offers pages for general interest, students, teachers (including access to curriculum-driven publications), ecological issues, and career information.

Earth's 911
http://www.earth911.org/master.asp
Aimed at children, students, and the general public, this site offers information on recycling and waste management options, including links to state-by-state listings of recycling centers, businesses, and regulations.

Ecological Society of America
http://www.esa.org
A nonpartisan, nonprofit organization of scientists founded in 1915 to "promote ecological science by improving communication among ecologists; raise the public's level of awareness of the importance of ecological science; increase the resources available for the conduct of ecological science; and ensure the appropriate use of ecological science in environ-

mental decision making by enhancing communication between the ecological community and policy-makers." Sponsors the Sustainable Biosphere Initiative. Offers an Education Section as well as resources for career placement, publications, and links to other ecology sites.

Ecology WWW Page
http://www.botany.net/Ecology
Maintained by Anthony R. Brach of the Missouri Botanical Garden and Harvard University Herbaria, this list of links is maintained as a volunteer service to students, teachers, researchers, and others interested in the science of ecology.

Envirolink
http://www.envirolink.org
This nonprofit organization has been providing access to online environmental resources since 1991, organized by categories such as Ecosystems, Environmental Disasters, Forests, and many more. The Ecology subsection files sites under such subcategories as Actions You Can Take, Educational Resources, General Info, Government Resources, Organizations, Jobs and Volunteer Opportunities, and Publications.

Environmental Protection Agency
http://www.epa.gov
This U.S. government agency offers information and Web links organized by ecosystems such as the following: aquatic ecosystems, coasts, coral reefs, deserts, ecological assessment, ecological monitoring, ecological restoration, endangered species, environmental indicators, estuaries, exotic species, forests, freshwater ecosystems, lakes, landscape ecology, marine ecosystems, oceans, species, terrestrial ecosystems, urban ecosystems, watersheds, and wetlands.

National Academy of Sciences
http://www4.nationalacademies.org/nas/nashome.nsf
A "private, nonprofit, self-perpetuating society of distinguished scholars engaged in scientific and engineering research, dedicated to the furtherance of science and technology and to their use for the general welfare. Upon the authority of the charter granted to it by the Congress in 1863, the Academy has a mandate that requires it to advise the federal government on scientific and technical matters."

National Council for Science and the Environment
http://www.ncseonline.org
A U.S. government-sponsored Web site which links to the National Library for the Environment. From here, students can search more than 1,200 congressional reports on topics in agriculture, biodiversity, climate change, energy, forests, pesticides, pollution, public lands, stratospheric ozone, wetlands, and a host of other environmental topics.

National Oceanic and Atmospheric Administration
http://www.noaa.gov
NOAA maintains pages on oceans, satellites, fisheries, weather, climate, coasts, and current events broken down by type of environmental impact: dust, fires, floods, ice, oil spills, snow, severe weather, tropical, and volcanic. In addition, NOAA's library, photo library, other pages on topics of ecological interest (such as coral reefs and El Niño), and organized lists of links provide access to other sites and research.

National Resources Conservation Service
http://www.nrcs.usda.gov
The mission of this division of the U.S. Department of Agriculture is to "assist owners of America's private land with conserving their soil, water, and other natural resources." The NRCS maintains two pages of interest: "Ecological Sciences" for agribusiness professionals and a page for teachers and kids with educational materials on soils, conservation biology, composting, and the like.

The Need to Know Library: Ecology and Environment Page
http://www.peak.org/~mageet/tkm/ecolenv.htm
Lists topics "under three headings: Ecology, Conservation Biology, and Botany and Systematics. Links to home pages of botanical and ecological societies are provided in the Journals and Societies List. The Ecological Software List offers links to sites describing software applications for analyzing or organizing ecological data. The Natural Resource Agencies List provides links to home pages of government agencies that conduct ecological research or that are responsible for the management of various ecosystems. The Natural Areas and Reserves heading lists links to sites that feature information about natural areas, LTER sites, national parks, and wilderness areas. The Environment List refers to sites dealing with ecosytem conditions and the Environmental Societies List is a catalog of links to home pages of environmental organizations."

Sierra Club

http://www.sierraclub.org

In addition to its outreach programs encouraging public involvement in recycling, conservation, and nature activities, the Sierra Club is active in environmental issues, offering updates on current events in water, energy, global population, and more.

Society for Conservation Biology

http://conbio.net

The Virtual Library of Ecology and Biodiversity is maintained through this site.

UNEP-WCMC Protected Areas Information

http://www.wcmc.org.uk/data/database/un_combo.html

Sponsored by the United Nations Environment Programme and the World Conservation, this page provides access to information on nearly all nations of the world, including basic facts, history, and present policies concerning protected areas. Each essay is lengthy and detailed and includes references.

United National Environment Programme

http://www.unep.org

UNEP's mission is "to provide leadership and encourage partnership in caring for the environment by inspiring, informing, and enabling nations and peoples to improve their quality of life without compromising that of future generations." Its Web site offers resources and links on habitats, species, regions, climate change, protected areas, and much more—including a kids' page that offers facts and educational games on the atmosphere, forests, biodiversity, polar regions, whales, and aquatic ecosystems.

U.S. Geological Survey

http://www.usgs.gov

Not simply a geology resource, the USGS maintains a wealth of data and fact sheets on biological and ecological information as well, searchable by topic or state. Topics include Biological Resources, Earthquakes, Floods, Ground Water, Maps, Streamflow, Water Quality, and more.

USDA Forest Service

http://www.fs.fed.us

This U.S. government site offers information on public lands in national forests and grasslands, organized state by state, as well as a photo gallery, maps, and information on the programs overseen by the agency.

World Conservation Union

http://www.iucn.org

Known also by its former initials IUCN, the World Conservation Union "seeks to influence, encourage and assist societies throughout the world to conserve the integrity and diversity of nature and to ensure that any use of natural resources is equitable and ecologically sustainable." Provides access to library catalog, publications, programs, contacts.

World Resources Institute

http://www.wri.org

A nonprofit environmental research and policy organization that works with governments, the private sector, and civil society groups in more than one hundred countries around the world. Offers updates on global environmental and ecological issues. Maintains a section called "EarthTrends" in which maps, data tables, searchable databases, and country profiles can be accessed by topic: Coastal and Marine Ecosystems; Water Resources and Freshwater Ecosystems; Climate and Atmosphere; Population; Health and Human Well-being; Economics; Business and the Environment; Energy and Resources; Biodiversity and Protected Areas; Agriculture and Food; Forests, Grasslands and Drylands.

CATEGORIZED INDEX

SUBJECT INDEX

Australia, habitat loss, 207
Australian honey eaters, 636
Autochthonous deposition, 465
Autotomy (defense mechanism), 129
Autotrophs, 50, 86, 111. *See also* Producers
Auxins, 650

Baboons; competition among, 114; hierarchies, 330; social systems, 386
Bacillus cereus (biopesticide), 66
Bacillus popilliae (biopesticide), 66
Bacillus thuringiensis (biopesticide), 66, 285
Backcrossing, 83
Background extinctions, 246
Bacteria; anaerobic, 224; bioluminescent, 43; biopesticides, 65; cyanobacteria, 225, 393, 465, 482; as decomposers, 112, 185, 290, 368; in digestion, 102, 305; mutation rates in, 438; nitrogen fixation, 289; oil-eating, 448; phosphorus cycle, 290; in soil, 316, 596; symbiotic, 127; weathering agents, 173
Bacterioplankton, 392-393
Badgers, defense mechanisms of, 126
Baer, Karl Ernst von, 161
Baer's laws, 161
Balance of nature, 28, 31, 255, 517
Bald eagles, 199
Barnacles, 623
Barrier reefs, 564
Base flow, 341
Batesian mimicry, 128, 416, 538
Bathypelagic marine zone, 392

Bats; desert, 157; as pollinators, 497
Beaded lizards, 486
Beagle, voyage of the, 238
Bears, 13
Beebe, William, 197
Beefalo, 84
Beer, Gavin de, 163
Bees, 344; Africanized, 116; dance language, 215; as pollinators, 496
Beetles; bark, 511; bombardier, 126; as pollinators, 497
Behavior; as camouflage, 74; and communication, 95; ethology, 215; and reproductive strategies, 576
Behavioral adaptations, 10
Behavioral displays, 167
Behavioral ecology, 218. *See also* Categorized Index
Behavioral genetics, 217
Behavioral isolation, 358
Behaviorism, 216
Benthic marine zone, 391, 397
Bergmann's rule, 80
Beta-carotene, 405
Beta individuals, 330
Bighorn sheep, 157
Bioaccumulation. *See* Biomagnification
Biodiversity, 32, 36, 134, 618; and adaptation, 7; ancient, 251; and biogeography, 37; and pollution, 500; rain forests, 558; urban and suburban wildlife, 660
Biogenetic law, 162
Biogeochemical cycles; hydrologic cycle, 338, 342; nutrient cycles, 440, 443. *See also* Carbon cycle;

Nitric oxides, 501
Nitrogen, excessive, 222
Nitrogen cycle, 185, 289, 441-442
"No net loss" policy, 584
NOAA. *See* National Oceanic and
 Atmospheric Administration
Nocturnal animals, 26, 158, 315,
 359, 488, 663
Nolen, Thomas G., 322
Nomadic animals, 407
Nonbranching evolution, 228
Nongeographic speciation, 605
Nonrandom mating, 435, 439
Nonsymbiotic mutualism, 25
North American biomes, 314
North American prairies, 299
North American taiga, 629, 632
North American Wetlands
 Conservation Act (1989), 675
North American Wetlands
 Conservation Reauthorization
 Act (2002), 675
Noss, Reed F., 34
Nuclear and radioactive waste,
 669
Nuclear weapons, 669
Numerical response of predators,
 536
Nutrient cycles, 185, 440, 443; soil
 generation, 596. *See also*
 Geochemical cycles
Nutrients, excess, 222, 226

Oak Ridge National Laboratory,
 180, 188
Ocean biomes, 63, 391. *See also*
 Marine biomes
Ocean pollution, 444, 451
Oceanic marine zone, 391
Octopuses, mimicry in, 417
Odum, Eugene P., 191

Offspring care; altruism, 19-20;
 camouflage, 74; territoriality,
 635
Oil deposits (tundra), 657
Oil spills, 444, 451
Old-growth forests, 133, 265, 452,
 454
Olfactory displays, 169
Oligotrophic lakes, 225
Omnivores, 112, 257, 455-456
*On the Origin of Species by Means of
 Natural Selection* (Darwin), 29,
 179, 215, 238, 274
Ontogeny, 161
Operation Desert Storm, 449
Operators (mimicry), 415
Opossums, 665
Optimality theory, 639
Opuntia (prickly pear cactus), 120
Orchidaceous mycorrhizae, 426
Orchids; endangered, 208;
 pollination, 498
Ordovician period, 90
Organismal ecology, defined, 171
Organophosphates, 471
Orians, G. H., 115
Orientation response, 321
Ornithischians, 158
Outcrossing, 286
Overcultivation, 150
Overgrazing, 150, 300, 304, 307,
 336, 561; southern Africa, 207
Overland flow, 340
Ovulation, 577
Owls; barn, 665; screech, 665;
 short-eared, 662; snowy, 662;
 spotted, 109, 203, 453
Oxygen cycle, 288; rain forests,
 555
Oxygen depletion, eutrophic
 environments, 223

Shells, as defense mechanisms, 125

Shells, sea, 465

Sherrington, Charles, 323

Sickle-cell disease, 520

Sierra Club, 123, 678

Sierra Nevada, 77

Signal pheromones, 476

Silent Spring (Carson), 30, 193, 200

Silurian period, 90, 92

Silverswords, 13

Similarity (for community classification), 109

Single larger or several smaller (SLOSS) controversy, 33, 379

Siphonophores, 45

Skunks, 665; defense mechanisms, 127

Slash-and-burn agriculture, 264, 272, 590, 593; and deforestation, 132; global warming, 295; reforestation, 572

Sleep, 321

Slobodkin, Lawrence B., 255

SLOSS (single larger or several smaller) controversy, 33, 379

Sludge treatment and disposal, 669

Smell; displays, 169; and pheromones, 478

Smith, Frederick E., 255

Snags, 452

Snail darters, 202

Snakes, venomous, 487

Snow geese, 435

Snowpack, 341

Snowshoe hares, 514

Social Darwinism, 234

Social ecology, 124

Social systems; insects, 95, 343, 350; mammals, 385, 390

Society of American Foresters, 268

Sociobiology, 218

Softwoods, 270

Soil, 594, 600; contamination, 601, 603; degradation, 572; erosion, 211, 214, 591; nutrients in, 440; sampling, 597; sterilants, 471; in waste management, 600; wetlands, 672. *See also* Erosion; Soil chemistry

Soil chemistry, 597

Soil conservation, 213

Soil Conservation Service, 677

Soil contamination, 131

Sokolov, E. N., 320, 322

Solar radiation, 69; greenhouse effect, 308

Solar tracking, 653

Solenopsis saevissima (fire ants), 345

Solid waste, 667

Sonoran Desert, 156

Source reduction, 670

South American dry forests, 588

Sparrows, dusky seaside, 203

Speciation, 239, 276, 604, 607; adaptive radiation, 12, 14; and isolating mechanisms, 362; punctuated equilibrium, 543, 548. *See also* Categorized Index

Species; definitions, 228; formation, 543; population fluctuations, 513; and territoriality, 633

Species diversity, 32

Species loss, 376, 608, 611; animals, 196, 204; plants, 205, 210. *See also* Extinctions

Species-removal studies, 609

Species selection, 545

Sphenophyta (horsetails), 241

Sphinctozoans, 568